Designing Aeration Systems using Baseline Mass Transfer Coefficients

For Water and Wastewater Treatment

T0199785

Johnny Lee MSc, MICE, CEng, P.Eng

CRC Press
Taylor & Francis Group
Boca Raton   London   New York

CRC Press is an imprint of the
Taylor & Francis Group, an **informa** business
A SCIENCE PUBLISHERS BOOK

Cover credit: Cover illustration reproduced by kind courtesy of the author.

First edition published 2021
by CRC Press
6000 Broken Sound Parkway NW, Suite 300, Boca Raton, FL 33487-2742

and by CRC Press
2 Park Square, Milton Park, Abingdon, Oxon, OX14 4RN

Library of Congress Cataloging-in-Publication Data

Names: Lee, Johnny, 1952- author.
Title: Designing aeration systems using baseline mass transfer coefficients
 : for water and wastewater treatment / Johnny Lee, American Society of
 Civil Engineers, The City of Waterloo (Service Centre), Waterloo,
 Ontario, Canada.
Description: First edition. | Boca Raton : CRC Press, Taylor & Francis
 Group, 2021. | Includes bibliographical references and index.
Identifiers: LCCN 2020055218 | ISBN 9780367617615 (hardcover)
Subjects: LCSH: Water--Aeration--Mathematical models. | Mass
 transfer--Mathematical models.
Classification: LCC TD458 .L44 2021 | DDC 628.1/65--dc23
LC record available at https://lccn.loc.gov/2020055218

ISBN: 978-0-367-61761-5 (hbk)
ISBN: 978-0-367-61764-6 (pbk)
ISBN: 978-1-003-10646-3 (ebk)

Typeset in Times
by TVH Scan

Preface

Nowadays, the way is totally open for an individual to revolutionize physics. Perhaps someone will devise a new interpretation of a measurement problem, or show us how several fundamental constants are really related. The same thing happens over and over again in various fields; existing theories become entrenched, elaborations of them become increasingly detailed and increasingly expensive, then someone produces a radically new theory that seems to come out of nowhere (but actually doesn't) and the cycle starts again.

Many people argue that people like Einstein and Newton took the low-hanging fruit and that it's no longer possible for an individual to make great advances working largely alone. I disagree with this view (as I have worked all alone for many years), even leaving aside the fact that Newton and Einstein relied heavily on the work of others, as I have relied on other people's findings and data. What they came up with were paradigm shifts: not merely elaborations of existing theories, but really new ways of looking at things. This happens very rarely. On the contrary, by their very nature, it's hard for teams to devise paradigm shifts. Teams have to set out plausible-sounding grant applications, ones where they can expect to make some useful progress. An individual, working in his or her spare time or within the protection of tenure (or in my case my own savings), can sometimes afford to devote a large amount of time to working on something that sounds a bit crazy but might actually be right.

While producing a theory that sounds crazy is no proof at all that the theory is correct (the test of that is superior agreement with experimental evidence), paradigm-shifting theories do always sound a bit crazy.

Newton faced criticism from people who objected that he had specified no mechanism for gravity, which he liked to call "the hand of God". Einstein faced opposition from people who were certain that time could not be relative and the luminiferous aether (a staple of physics at the time, much talked about by the Michio Kakus of the day) must clearly exist. (*I am facing opposition because many people are certain that the mass transfer equation cannot contain double the amount of respiration rate R, resulting in a mass transfer coefficient that is twice the value or so that would be calculated from the traditional equation for the steady-state test. This is briefly explained in Chapter 1).*

Neither of them would have been taken seriously had proof of their theories not been forthcoming. (*My published papers [Lee 2018, 2019a, b] contain case studies of previous works by other researchers and mathematical proof of my theories---all it needs is some testing to confirm and hence grants and funding are required*).

My new theory is based on the concept of a baseline for the mass transfer coefficient. As it has never been used before, the discovery of this concept for designing aeration systems is truly ground-breaking. But what is a baseline? *A baseline is a fixed point of reference that is used for comparison purposes.* The baseline serves as the starting point against which all future estimations are measured. A baseline can be any number that serves as a reasonable and defined starting point for comparison purposes. It may be used to evaluate the effects of a change, track the progress of an improvement project, or measure the difference between two periods of time. One of the usefulness of a baseline mass transfer coefficient, in the context of wastewater treatment, is to predict what will happen to the oxygen transfer in aeration systems if microbial activity is present in the test water.

In line with Mahendraker's theory (Mahendraker 2003), the author believes that the resistance to oxygen transfer is composed of two parts: one, a resistance by the reactor solution; the other is the resistance due to the biological floc (Mahendraker et al. 2005a, 2005b). So, if clean water testing is considered a baseline, the effect of cell concentrations can be superimposed onto the baseline to determine the oxygen transfer in the actual mixed liquor of a treatment plant aeration basin, using the Principle of Superposition in physics. This concept has been criticized by some experts. One university professor advised me: "Consulting companies cannot benefit from investing in this research area. The only sector that I can think of which may have an interest may be those that make aeration devices. However, they again would have to see a benefit from investing in your research. So, currently, they would do standardized testing and have field data to verify their design calculations. Can investing in your research help them improve their design or competitiveness leading to possible economic benefits? If so, they are the ones you may follow up with".

"The other problem that I see is the cost and practicality of your research. Real life testing is expensive enough and requires an expensive infrastructure. I don't see anyone funding that kind of a setup only for this study. The facilities that offer services for standard water testing – I don't see them allowing the use to do testing with wastewater. It may be worth examining if there is a simpler and cheaper way of getting some data that will allow you to "validate" your model. Lower cost of the study might make it easier to find an industrial partner. However, again, they would want to see a value to them from investing in the research".

This raises the question "who is responsible for each aspect of aeration systems design?" According to Stenstrom and Boyle (1998): "Some owners and consultants want to make it entirely the responsibility of the manufacturer. Such attitudes are incorrect and produce indifferent attitudes during the design process ("it is their job-why are we worrying about it?") and wasteful litigation. Manufacturers are held responsible for the aspects of aeration design that they control — this means clean-water transfer performances and mechanical integrity. However, manufacturers have no control over wastewater characteristics, process design, or the way their equipment is operated and maintained. Design engineers must anticipate a range of operating conditions and their effects on the aeration system. Alpha factors (RATIOS OF THE TRANSFER COEFFICIENTS between tap water and wastewater) are strongly affected by process design and operation and by system configuration, as well as wastewater characteristics." [*In this respect, my finding differs from conventional thinking: in my opinion, alpha should only be dependent on the reactor solution, not the process, in line with Mahendraker's (2003) thinking*].

Another criticism comes from the American Oxygen Transfer Standards Committee: "You're confusing oxygen transfer mechanisms with respiration. The respiration determines the oxygen uptake rate (OUR) which is then matched by the oxygen transfer rate. There is no double R term. R is NOT a part of the oxygen transfer mechanism".

Stenstrom and Boyle (1998) continued: "Currently, Alpha factors cannot be specified by the manufacturer: they must be determined by the design engineer for the range of operating conditions anticipated by the owner. Alpha factors for design are not a single value but ranges of values that occur for different process conditions, times, and locations within aeration tanks. Owners must know and accept responsibility for their operating decisions; for example, dramatically reducing sludge age decreases oxygen transfer efficiency in most aeration systems, which should be considered before making process changes. Information on alpha must be obtained through in-process testing experience or carefully documented data from the literature or other credible sources. Small-scale testing such as laboratory testing has not been a reliable source of data." [*Here again, my proposed model appears to be able to utilize such data to predict full-scale performance*].

"Both owners and designers should not accept alpha factor claims by manufacturers. In most cases, it will be very difficult to hold manufacturers accountable for the alpha factors they might claim, because they cannot control process operation or wastewater characteristics. Consultants who accept manufacturers' recommendations without verifying them are not protecting their clients. Another important issue that remains the responsibility of the design engineer is the compliance specification of the aeration equipment and the provision of any required scale-up to the actual installation. Typically in the United States, clean water compliance specifications are used. For process water, compliance specifications require considerations of wastewater variability and process loading that lead to substantial uncertainty. If left to the manufacturer, extremely conservative and costly systems will result for obvious reasons, because the manufacturer has little knowledge of the wastewater and process operating conditions. In this situation, it is incumbent on the designer to specify alpha and other process variables within the specification. Another issue in compliance testing is scale-up. If shop tests are to be performed, it is up to the designer to specify the shop test and to provide the necessary scale-up to field conditions. Manufacturers may be consulted on issues, but it is ultimately the designers' responsibility to ensure proper scale-up. To avoid this problem, some designers specify clean water compliance testing in the field system".

From the discussion above, it would appear that approaching aeration device manufacturers may not be a fruitful outcome. Maybe industrial partners such as paper and pulp companies will be interested. Does anybody know of any? But my first choice would still be government bodies and academic organizations and institutions, I reckon.

REFERENCES

Lee, J. (2018). Development of a model to determine the baseline mass transfer coefficients in aeration tanks. Water Environment Research 90(12): 2126.

Lee, J. (2019a). Baseline mass transfer coefficient and interpretation of non-steady state submerged bubble oxygen transfer data. 10.1061/(ASCE)EE.1943-7870.0001624.

Lee, J. (2019b). Is Oxygen Transfer Rate (OTR) in Submerged Bubble Aeration affected by the Oxygen Uptake Rate (OUR)? 10.1061/(ASCE)EE.1943-7870.0001635.

Mahendraker, V. (2003). Development of a unified theory of oxygen transfer in activated sludge processes – the concept of net respiration rate flux. Department of Civil Engineering, University of British Columbia.

Mahendraker, V., Mavinic, D.S., Rabinowitz, B. (2005a). Comparison of oxygen transfer test parameters from four testing methods in three activated sludge processes. Water Quality Research Journal Canada, 40(2): 164-176.

Mahendraker, V., Mavinic, D.S., Rabinowitz, B. (2005b). A simple method to estimate the contribution of biological floc and reactor-solution to mass transfer of oxygen in activated sludge processes. Wiley Periodicals, Inc. DOI: 10.1002/bit.20515.

Stenstrom, M.K., Boyle, W. (1998). Aeration systems-responsibilities of manufacturer, designer, and owner. Environmental Engineering Forum, Journal of Environmental Engineering, May 1998 (398).

Contents

1

Prologue

The US EPA in the 70s poured in substantial amount of money to fund fundamental research as they recognized the importance of the connection between clean water tests and wastewater tests. Although they have made substantial progress, the fundamental question of relating clean water and wastewater tests remains unresolved [Mahendraker et al. 2005]. A new revolutionary finding may revive their interest.

This book is focused primarily on submerged bubble aeration. In aeration systems, diffused air is a simple concept which entails pumping (injecting) air through a pipe or tubing and releasing this air through a diffuser below the water's surface. The submerged system has little visible pattern on the surface, and is able to operate in depths up to and exceeding 12 m (40 ft). The best aerators use quiet on-shore compressors that pump air to diffusers placed at a pond or tank bottom. From stone diffusers to self-cleaning dome diffusers, they release oxygen throughout the water column creating mass circulation that mixes bottom and top water layers, breaks up thermal stratification, and replenishes dissolved oxygen through molecular oxygen mass transfer by means of gas diffusion. Gas transfer is the exchange of gases between aqueous and gaseous phases. In a diffused aeration, gas exchange takes place at the interface between submerged air bubbles and their surrounding water. According to Lewis and Whitman (1924), these bubbles are each wrapped with two layers of films through which the gas must go through. The transfer rate is usually expressed by a mass transfer coefficient symbolized by $K_L a$.

No one has seen the two films around a bubble, let alone measuring the thicknesses of these films based on which $K_L a$ can be quantified. The coefficient can only be determined by an indirect method, such as the one used by the current ASCE standard (ASCE 2007). The transfer rate can also be determined by mass balances-the gas depletion rate from the bubble must equal the oxygen uptake rate in the liquid. This concept of gas-side oxygen depletion is not as readily understood as it may seem:

The respiration determines Oxygen Uptake Rate (OUR) that equates to the Oxygen Transfer Rate (OTR) at steady-state. The understanding that "the respiration determines the OUR which is then matched by the oxygen transfer rate" concurs with my thesis in this book, and indeed is correct. But in submerged aeration, there is the phenomenon known as gas-side oxygen depletion, so that the oxygen transfer rate is affected by this effect and this effect (incorrectly) changes the value of $K_L a$. To make the correction, the OTR is therefore given by $K_L a_f (C_{\infty f}^* - c) - R$ (where f means "under field conditions") under the principle of superposition in physics (this concept is further explained in Chapter 6). This is then matched by the oxygen transfer rate OTR at steady-state, which is equal to respiration rate R.

Therefore,

$$K_L a_f (C^*_{\infty f} - c) - R = R$$

Although R is not part of the oxygen transfer mechanism, gas-side oxygen depletion is. If R is non-variant within the test period, then it can be determined in a gas flow steady state, where R is matched by the gas depletion rate in the bubbles which affects the value of the OTR. This has led to the above equation where $K_L a_f$ is understood to be (alpha.$K_L a$), where alpha is a function of the wastewater characteristics only. The current alpha as used in the conventional model treats it as a lumped parameter that envelops both effects (water characteristics and gas depletion), making it a highly variable parameter that is indeterminate. The concept of gas-side depletion of oxygen from air bubbles, at first glance, appears to be simple and straightforward, but is in fact less readily understood than it may seem. In ordinary air bubble aeration, the OTE is typically 10 ~ 20%, since oxygen gas is only slightly soluble in water. (In clean water, it can be as much as 40% depending on the aeration device and the mixing intensity). This 10 ~ 20% by weight is the actual amount of oxygen successfully being transferred to the liquid. This quantity is exactly equal to the quantity of gas depleted from the air bubbles, since 'oxygen transfer' and 'gas-side gas depletion' are one and the same.

In fact, gas-side gas depletion is the only means of oxygen transfer for bubbles rising to the water surface, when any other means of transfer such as the free water surface or the bubble formation at the diffuser are negligible. Therefore, any modeling of oxygen transfer into any liquid (tap water, sewage, industrial wastes, etc.) must include the gas depletion effect, otherwise, the model cannot be valid.

In the review paper by Uby (2019), in section 6.1, it was stated: "Among the CEN and DWA standard test methods, no result dependence on initial conditions (supersaturation or depletion of oxygen and nitrogen) has been detected (Wagner and Popel 1996), but a rigorous uncertainty analysis is lacking. This has been fully accepted in German and European practice (DWA 2007, CEN 2003). Gas side depletion of oxygen from air bubbles has been shown to be a minor concern under common conditions (Brown and Baillod 1982, Jiang and Stenstrom 2012), corroborating this approach [of Lars' paper]. In the interest of standardization and uncertainty quantification, the difference should be quantified, though experience speaks for at most a minor impact."

In my opinion, the above understanding by Lars (as well as the various standards) is absolutely incorrect. It should be taken only in the context of the results of a clean water test, where the parameters C_S (saturation value) and $K_L a$ (mass transfer coefficient) are estimated. The reason why no result dependence on initial conditions (supersaturation or depletion of oxygen and nitrogen) has been detected is because these effects have already been absorbed in the Standard Model. In other words, the calculated results of the two parameters have already included any dependence on these effects, even though such dependence is not detected. In the application of clean water results to sewage or other liquid, these effects will change in accordance with changes in the gas depletion rate which is the same as changes in oxygen transfer rate under a changed environment. The bacterial and other microbial composition and their metabolic functioning, in particular, constitute drastic changes in the oxygen gas depletion rate which then drastically *affects the value of the gas transfer parameters, in particular the mass transfer coefficient $K_L a$. Even if all means of standardization and uncertainty analyses are carried out, they will not significantly improve the clean water test results. On the other hand, if clean water test results are to be translated to other fluids with significant microbial cell content, mixed liquor for example, then the principle of superposition must be applied to the Standard Model to take into account this all-important gas depletion effect in diffused aeration, without which nothing in terms of oxygen transfer happens. Therefore, gas-side gas depletion is not a minor impact, but in fact is the ONLY impact in submerged diffused-air bubble aeration, the magnitude of which is a function of a myriad of variables.*

The amount of gas depleted from the bubble at any time not only depends on the films, but also on the path taken by the bubble that follows a gas depletion curve which is a function of many variables. This curve would vary with different heights and depths. Also, this gas depletion curve in clean water is substantially different from that in wastewater. The loss rate of gas from the bubble is the gain rate of transfer in the case of clean water aeration at any time and place inside an aeration tank.

Given that the mass transfer coefficient $(K_L a)$ is a function of many variables, in order to have a unified test result, it is necessary to create a *baseline* mass transfer coefficient, so that all tests will have the same measured baseline. $K_L a$ is found to be an exponential function of this new coefficient and is dependent on the height of the liquid column (Z_d) through which the gas flow stream passes. DeMoyer et al. (2003) and Schierholz et al. (2006) have conducted experiments to show the effect of free surface transfer on diffused aeration systems, and it was shown that high surface-transfer coefficients exist above the bubble plumes, especially when the air discharge (Q_a) is large. When coupled with large surface cross-sectional area and/ or shallow depth, the oxygen transfer mechanism becomes more akin to surface aeration where water entrainment with air from the atmosphere becomes important. The water turbulence has a significant effect on oxygen transfer. The alternative to a judicious choice of tank geometry and/or gas discharge is perhaps another mathematical model that could separate the effect of surface aeration from the actual aeration under testing in the estimation of the mass transfer coefficient. This separate modelling for surface aeration is not a topic in this book. Nevertheless, a simple graphical method to take this effect into account in the establishing of the *baseline* coefficient is proposed in Chapter 6, Section 6.6.4.

In engineering, the mass transfer coefficient is a diffusion rate constant that relates the mass transfer rate, mass transfer area, and concentration change as driving force, using the Standard Model, typically stated in the form given by Eq. 4.1 in Chapter 4. This can be used to quantify the mass transfer between phases, immiscible and partially miscible fluid mixtures (or between a fluid and a porous solid). Quantifying mass transfer allows for design and manufacture of separation process equipment that can meet specified requirements and estimate what will happen in real life situations (chemical spill, wastewater treatment, fermentation, and so forth) if the effect of other factors, such as turbulence either due to the free surface exchange or due to mechanical mixing within the water body, can be isolated or eliminated or modelled separately.

Mass transfer coefficients can be estimated from many different theoretical equations, correlations, and analogies that are functions of material properties, intensive properties and flow regime (laminar or turbulent flow), all based on the Standard Model. Selection of the most applicable model is dependent on the materials and the system, or environment, being studied. This book is about the discovery of a new coefficient called the *baseline* mass transfer coefficient (K_La_0). The process of this discovery is described in Chapter 3. For open tank aeration, the author defines it as the ordinary mass transfer coefficient (K_La) measured at the **equilibrium** pressure of the standard sea-level atmospheric pressure (101.325 kPa). Since most testing is carried out in a vessel of some physical height, the equilibrium pressure must exceed this baseline pressure of 1 atmosphere. If water is used for an aeration test in accordance with current standards [ASCE 2007] [CEN 2003] [DWA 2007], the system would attain a "super-saturated" state at equilibrium. This super-saturated dissolved gas concentration (C_∞^*) would differ from the saturation concentration that can be readily found from published data or any chemistry handbook on gas solubility. The closest experiment that would yield a handbook solubility (C_S) value (and the corresponding baseline mass transfer coefficient) would be a laboratory-scale experiment.

In any other situations, K_La_0 is not directly measurable. This book is about how the baseline can be determined using the Standard Model for gas transfer, despite the many variables affecting such transfer and K_La. Based on the various literature data cited, the baseline has proven to be a valuable parameter (perhaps even more useful than K_La itself) that can be used to predict gas transfer under different test conditions, such as different heights or liquid depths, perhaps even different geometry. This is a *revolutionary change* as, up to now, it has not been possible to correlate K_La from one test to another, even under ordinary testing circumstances. However, the baseline, or more correctly the specific baseline upon normalizing with the gas flowrate, is a "true" constant for every test. In the context of the meaning of "baseline", the book is expected to be a baseline itself for future upgrading when more data becomes available. People interested in this book would certainly be scientists, engineers, researchers, treatment plant operators, and manufacturers of aeration systems.

As mentioned, the mass transfer coefficient K_La is related to the air discharge and is found to be dependent on the gas average flow rate (Q_a) passing through the liquid column. Q_a is estimated from the gas *mass* flow rate (Q_S), and is expressed in terms of *actual* volume of gas per unit time, as distinct from Q_S that is expressed as *mass* per unit time. For a uniform liquid temperature (T) throughout the liquid column, Q_a is calculated by Boyle's Law, and taking the arithmetic mean of the volumetric flow rates over the tank column. (Although the mass flow rate Q_S is sometimes also expressed as volume per unit time, it is not true volume because it is expressed as standard conditions, which is equivalent to mass per unit time). As such, Q_a is a variable dependent on temperature, pressure and volume, even when the gas supply Q_S is fixed and non-variant.

When an intensive property such as temperature is varied, K_La_0 is directly proportional to this mean gas flow rate (Q_a) to a power q, where q is usually less than unity for water in a fixed column height and a fixed gas supply rate at standard conditions, (Q_S). However, K_La_0 is not proportional to Q_S, as the case studies presented in this book would show, although there may be good correlation in some cases. When temperature is fixed, the same relationship holds for different values of Q_a, regardless of column height. This book provides theoretical development and case studies that verify this baseline which can be standardized specifically to the average gas flow rate as a new function $(K_L a_0)/Q_a^q$ that is applicable to submerged bubble aeration testing. This function is termed the *specific baseline* in this document, and is a constant quantity for any test temperature T. This relationship between the baseline and gas flow can be determined by experiments, as the case studies in Chapter 5 demonstrate, in which it is shown that the overhead (or headspace) pressure is also an intensive property that, when varied, would give the same baseline versus gas flowrate relationship. When the function is determined at standard conditions, it is termed the *standard specific baseline* expressed as $(K_L a_0)_{20}/Q_{a20}^q$ and is a constant independent of tank height Z_d and gas flow Q_a.

Lastly, the suggested replacement of the temperature correction model for the mass transfer coefficient that is based on the Arrhenius equation as stipulated by ASCE Standard 2-06 [ASCE 2007], with the new 5th power model (see Chapter 2), may be controversial because the former equation is well known and is being used by the standard for a long time. This controversy is not too important in this manuscript as all the tests cited were conducted in the neighborhood of 20°C (within the range of 10°C to 30°C), and so there are only small differences in the calculations of $(K_L a)_{20}$ or $(K_L a_0)_{20}$ between the two models. Nevertheless, a discussion is in order since the new model gives a slightly better correlation between the standard baseline and the gas flow rates in all cases. As Dr. Stenstrom explained for the background: "In the first version of the standard, we debated the value of theta (θ) ...and found that most of the literature data supported 1.020 to 1.028 with the diffused systems clustering toward the bottom of the range and the surface [aeration systems] clustering toward the top of the range" [Stenstrom and Lee 2014].

From this, it can be inferred that there may be two different ranges of the temperature correction factor θ for the two aeration systems referred to by Stenstrom. Based on analyzing literature data on diffused systems, the author found that the 5th power model fits more closely with a theta (θ) value in the range of 1.016~1.018 [Lee 2017] [Chapter 2] which is closer to the range for diffused systems. Furthermore, the 'standard-recommended' theta value of 1.024 is probably based on tests on conventional treatment plants or shop tests of similar height that is usually around 3 m (10 ft) to 4.5 m (15 ft). The 5th power model is suitable for 'zero' height since most laboratory tests were carried out on a benchscale of very little height. Since the baseline pertains to a mass transfer coefficient of an infinitesimally shallow tank, it would appear that this new 5th power model is more suitable for correcting the *baseline* to the standard temperature. It should be noted in passing that temperature is an intensive property (i.e. independent of scale), whereas $K_L a$ is a function of height and other variables, and it is dependent on scale; so, it cannot be accurately corrected by a single factor that summarily ignores changes in height and other factors.

The book is divided into eight chapters. Chapter 2 below deals with the derivation of the 5th power model for temperature correction. Chapter 3 deals with the development of the model to determine the baseline mass transfer coefficients in aeration tanks. Chapter 4 is dedicated to the derivation and theoretical development of the Lee-Baillod model on which the subsequent depth correction model is based. Chapter 5 illustrates the functionality of the Baseline Mass Transfer Coefficient and Interpretation of Non-steady State Submerged Bubble Oxygen Transfer Data. Chapter 6 asks the question if Oxygen Transfer Rate (OTR) in Submerged Bubble Aeration is affected by the Oxygen Uptake Rate (OUR), concerning the use of the baseline for in-process field working conditions? Chapter 7 recommends further research to elucidate the question posed in Chapter 6, and Chapter 8 is the epilogue that summarizes all the core findings. It is expected that this book would serve practitioners in the designing of aeration systems, as well as serve as Standard Guidelines for water and wastewater (both in-process and non in-process) oxygen transfer testing, enhancing the current standards and guidelines, ASCE 2-06 [ASCE 2007] and ASCE-18-96 [ASCE 1997].

REFERENCES

ASCE-2-06. (2007). Measurement of Oxygen Transfer in Clean Water. Standards ASCE/EWRI. ISBN-10: 0-7844-0848-3, TD458.M42 2007.

ASCE-18-96. (1997). Standard Guidelines for In-Process Oxygen Transfer Testing. ASCE Standard. ISBN-0-7844-0114-4, TD758.S73.

Brown, L.C., Baillod, C.R. (1982). Modeling and interpreting oxygen transfer data. ASCE, Journal of the Environmental Engineering Division 108(EE4): 607.

CEN. (2003). EN 12255-15. Wastewater Treatment Plants – Part 15: Measurement of the Oxygen Transfer in Clean Water in Aeration Tanks of Activated Sludge Plants. European standard.

DeMoyer, C.D., Schierholz, E.L., Gulliver, J.S., Wilhelms, S.C. (2003). Impact of bubble and free surface oxygen transfer on diffused aeration systems. Water Research 37(8): 1890-1904.

DWA. (2007). Merkblatt DWA-M 209. Messung der Sauerstoffzufuhr von Beluftung-Seinrichtungen in Belebensanlagen in Reinwasser und in BelebtemSchlamm. (Formerly ATV; Measurement of oxygen transfer of aeration equipment in biological treatment plants in clean water and in activated sludge).

Jiang, Pan, Stenstrom, Michael K. (2012). Oxygen transfer parameter estimation: impact of methodology. Journal of Environmental Engineering 138(2): 137e142.

Lee, J. (2017). Development of a model to determine mass transfer coefficient and oxygen solubility in bioreactors. Heliyon 3(2): e00248.

Lewis, W.K., Whitman, W.G. (1924). Principles of gas absorption. Industrial and Engineering Chemistry 16(12): 1215-1220. Publication Date: December 1924 (Article) DOI: 10.1021/ie50180a002.

Mahendraker, V., Mavinic, D.S., Rabinowitz, B. (2005). Comparison of oxygen transfer parameters from four testing methods in three activated sludge processes. Water Quality Research Journal 40(2): 164-176.

Schierholz, E.L., Gulliver, J.S., Wilhelms, S.C., Henneman, H.E. (2006). Gas transfer from air diffusers. Water Research 40(5): 1018-1026.

Stenstrom, M., Lee, J. (2014). Private communication, email dated 2/7/2014.

Uby, Lars. (2019). Next steps in clean water oxygen transfer testing — A critical review of current standards. Water Research 157: 415-434.

Wagner, M.R., Popel, H.J., Johannes, H. (1996). Surface active agents and their influence on oxygen transfer. Water Science and Technology 34(3-4): 249-256. ISSN 0273-1223. https://doi.org/10.1016/0273-1223(96)00580-X.

2

Mass Transfer Coefficient and Gas Solubility

2.0 INTRODUCTION

The main objective of this chapter [Lee 2017] is to develop a mechanistic model (based on experimental results of two researchers, Hunter [1979] and Vogelaar et al. [2000], to replace the current empirical model in the evaluation of the standardized mass transfer coefficient $(K_L a)_{20}$ being used by the ASCE Standard 2-06 [ASCE 2007]. The topic is about gas transfer in water (how much and how fast), in response to changes in water temperature. This topic is important in wastewater treatment, fermentation, and other types of bioreactors. The capacity to absorb gas into liquid is usually expressed as solubility, C_S, whereas the mass transfer coefficient represents the speed of transfer, $K_L a$ (in addition to the concentration gradient between the gas phase and the liquid phase which is not discussed here). These two factors, capacity and speed, are related and the manuscript advocates the hypothesis that they are inversely proportional to each other, i.e. the higher the water temperature, the faster the transfer rate, but at the same time less gas will be transferred.

This hypothesis was difficult to prove because there is not enough literature or experimental data to support it. Some data [ASCE 1997] do support it, but they are approximate, because some other factors skew the relationship, for example, concentration gradient, and the hypothesis is only correct if these other factors are normalized or held constant.

This hypothesis may or may not be proved by theoretical principles, such as by means of thermodynamic principles to find a relationship between equilibrium-concentration and mass transfer coefficient, but such proof is beyond the expertise of the author. However, the hypothesis can in fact be verified indirectly by means of experimental data that were originally used to find the effects of temperature on these two parameters, solubility (C_S) and mass transfer coefficient $(K_L a)$. Temperature affects both equilibrium values for oxygen concentration and the rate at which transfer occurs. Equilibrium concentration values (C_S) have been established for water over a range of temperature and salinity values, but similar work for the rate coefficient is less abundant.

This chapter uses the limited data available in the literature to formulate a practical model for calculating the standardized mass transfer coefficient at 20°C. The work proceeds with general formulation of the model and its model validation using the reported experimental data. It is hoped that this new model can give a better estimate of $(K_L a)_{20}$ than the current method.

2.1   THE TEMPERATURE CORRECTION MODEL FOR $K_L a$

2.1.1   Basis for Model Development

Using the experimental data collected by two investigators, Hunter [1979] and Vogelaar et al. [2000], data interpretation and analyses allowed the development of a mathematical model that related $K_L a$ to temperature, advanced in this paper as a temperature correction model for $K_L a$. The new model is given as:

$$(K_L a)_T = K \times T^5 \times \frac{E \rho \sigma}{P_s} \qquad (2.1)$$

where $K_L a$ = overall mass transfer coefficient (min^{-1}); T = absolute temperature of liquid under testing in Kelvin; the subscript T in the first term indicates $K_L a$ at the temperature of the liquid at testing; and K = proportionality constant; E = modulus of elasticity of water at temperature T, (kNm^{-2}); ρ = density of water at temperature T, (kg m^{-3}); σ = interfacial surface tension of water at temperature T, (N m^{-1}); P_s is the saturation pressure at the equilibrium position (atm). The derivation is based on the following findings as described in Section 2.3.

The model was based on the two-film theory by Lewis and Whitman [1924], and the subsequent experimental data by Haslam et al. [1924], whose finding was that the transfer coefficient is proportional to the 4th power of temperature. Further studies by the subsequent researchers [Hunter 1979, Boogerd et al. 1990, Vogelaar et al. 2000] unveiled more relationships, which when further analyzed by the author, resulted in a logical mathematical model that related the transfer coefficient (how fast the gas is transferring when air is injected into the water) to the 5th order of temperature. Perhaps this is also a hypothesis, but it matches all the published data sourced from literature.

Similarly, using the experimental data already published for saturation dissolved oxygen concentrations, such as the USGS (United States Geological Survey) tables [Rounds 2011], Benson and Krause's stochastic model [Benson and Krause 1984], etc., it was found that solubility also bears a 5th order relationship with temperature.

So, there are actually three hypotheses. But are they hypotheses or are they in fact physical laws that are beyond proof? For example, how does one prove Newton's law? How does one prove Boyle's law, Charles' law, or the Gay-Lussac's law? They can be verified of course, but do not lend themselves easily to mathematical derivation using basic principles. As mentioned, Prof. Haslam found that the liquid film transfer coefficient varies with the 4th power of temperature, but how does one prove it by first principles? The model just fits all the data that one can find, although it would be great if it can be proven theoretically. However, the correlation coefficients for (Eq. 2.1) are excellent as can be seen in the following sections.

The paper for this chapter is not a theory/modelling paper in the sense that a theory was not derived based on first principles. Nor in fact is it an experimental/ empirical paper since the author did not perform any experiments. However, the research workers who did the experiments did not recognize the correlation, and so they have missed the connection. This paper revealed that these data can in fact support a new model that relates gas transfer rate to temperature that they missed. They used their data for other purposes, and drew conclusions for their purposes.

Further tests may therefore be required to justify these hypotheses. Although other people's data are accurate since they come from reputable sources, they are different from experiments specifically designed for this model development purpose only. The novelty of the proposed model is that it does not depend on a pre-determined value of theta (θ) in applying a temperature correction to $K_L a$ for a test data, if all other conditions affecting its value are held constant or convertible to standard conditions.

The current model adopted by ASCE 2-06 is based on historical data and is given by the following expression:

$$\frac{(K_L a)_{20}}{(K_L a)_T} = 1.024^{(20-T)} \tag{2.2}$$

In this equation, T is expressed in °C and not in K (Kelvin) defined for (Eq. 2.1). It has been widely reported that this equation is not accurate, especially for temperatures above 20°C. Current ASCE 2-06 employs the use of a theta (θ) correction factor to adjust the test result for the mass transfer coefficient to a standard temperature and pressure. The ratio of $(K_L a)_T$ and $(K_L a)_{20}$ is known as the dimensionless water temperature correction factor N, so that

$$N = \frac{(K_L a)_{20}}{(K_L a)_T} \tag{2.3}$$

Current model is therefore given by:

$$N = \theta^{(20-T)} \tag{2.4}$$

where θ is the dimensionless temperature coefficient. This coefficient is based on historical testing, and is purely empirical. Furthermore, the above equations indicate that the $K_L a$ water temperature correction factor N is exclusively dependent on water temperature. This is definitely not the case, as the correction factor is also dependent on turbulence, as well as other properties as shown in (Eq. 2.1). Current wisdom is to assign different values of theta (θ) to suit different experimental testing. While adjusting the theta (θ) value for different temperatures may eventually fit all the data, this may lead to controversies. Furthermore, it is necessarily limited to a prescribed small range of testing temperatures.

2.1.2 Description of Proposed Model

The purpose of this chapter is to improve the temperature correction method for $K_L a$ (the mass transfer coefficient) used on ASCE Standard 2-06 and to replace the current standard model by (Eq. 2.1).

The proposed model can also be expressed in terms of viscosity as described below. Viscosity can be correlated to solubility. When a plot of oxygen solubility in water is made against viscosity of water, a straight-line plot through the origin is obtained [IAPWS 2008]. When the inverse of viscosity (fluidity) is plotted against the fourth power of temperature, the linear curve as shown in Figure 2.1 below was obtained.

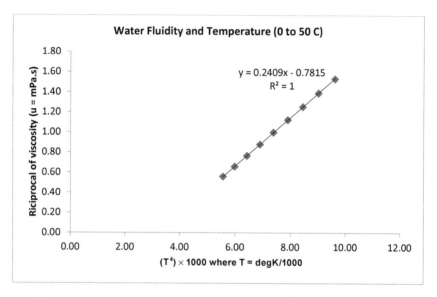

FIGURE 2.1 Reciprocal of viscosity plotted against 4th power of temperature

Therefore, viscosity happens to have a 4th order relationship with temperature, so that (Eq. 2.1) can be expressed in terms of viscosity and a first order of temperature, instead of using the 5th order term. The concept of molecular attraction between molecules of water and the oxygen molecule is important since changes in the degree of attraction would influence the equilibrium state of oxygen saturation in the water system as well as its gas transfer rate. Although the above plot (Figure 2.1) shows that the reciprocal of viscosity (fluidity) is linearly proportional to the 4th order of absolute temperature, the line does not pass through the origin.

As viscosity is closely correlated to solubility, it is obvious that the molecular attraction between water molecules that influences viscosity and the molecular attraction between water and oxygen molecules are interrelated. This correlation does not establish that an alteration of water viscosity, such as changes in the characteristics of the liquid, will have an impact on oxygen solubility. However, it will certainly affect the mass transfer coefficient. Viscosity due to changes in temperature is therefore an intensive property of the system, whereas viscosity due to changes in the quality of water characteristics is an extensive property. The equation relating viscosity to temperature is given by Fig. 2.1 as:

$$\frac{1}{\mu} = 0.2409 \times 10^3 \times \left(\frac{T}{1000}\right)^4 - 0.7815 \qquad (2.5)$$

where μ = viscosity of water at temperature T, (mPa.s)

Rearranging the above equation, T^4 can be expressed in terms of viscosity and therefore,

$$T^4 = K' \times \left(\frac{1}{\mu} + 0.7815 \right)$$ (2.6)

where K' is a proportionality constant.

Substitute (Eq. 2.6) into (Eq. 2.1), therefore,

$$(K_L a)_T = K \times \frac{(E\rho\sigma)_T}{P_S} \times K' \times \left(\frac{1}{\mu} + 0.7815 \right) \times T$$ (2.7)

Grouping the constants therefore,

$$(K_L a)_T = K'' \times \frac{(E\rho\sigma)_T}{P_S} \left(\frac{1}{\mu} + 0.7815 \right) \times T$$ (2.8)

where K'' is another proportionality constant.

Therefore, $K_L a$ can be expressed as either (Eq. 2.8) or as (Eq. 2.1). For the sake of easy referencing to this model, this model shall be called the 5th power model.

2.1.3 Background

The universal understanding is that the mass transfer coefficient is more related to diffusivity and its temperature dependence at a fundamental level on a microscopic scale. Although Lewis and Whitman long ago advanced the two-film theory [Lewis and Whitman 1924] and subsequent research postulated that the liquid film thickness is related to the fourth power of temperature in K [Haslam et al. 1924], it was not thought that this relationship could be applied on a macro scale. In a laboratory scale, Professor Haslam conducted an experiment to examine the transfer coefficients in an apparatus, using sulphur dioxide and ammonia as the test solute. Based on Lewis and Whitman's finding that the molecular diffusivities of all solutes are identical, he derived four general equations that link the various parameters affecting the transfer coefficients which are dependent upon gas velocity, temperature, and the solute gas. He found that the absolute temperature has a vastly different effect upon the two individual film coefficients. The gas film coefficient decreases as the 1.4^{th} power of absolute temperature, whereas the liquid film coefficient increases as the fourth power of temperature. The discovery that the power relationship between the liquid film coefficient and temperature can be applied to an even higher macroscopic level where C_S is a function of depth, is based on a combination of seemingly unrelated events as follows:

- Lee [1978] and Baillod [1979] derived, by theoretical and mathematical development, a formula for the mass transfer coefficient $(K_L a)$ on a macro scale for a bulk liquid treating the saturation concentration C_S as a dependent variable;
- The derived $K_L a$ mathematically relates to the "apparent $K_L a$" [ASCE 2007] that is defined in ASCE 2-06 standard;

- It was thought that K_L (the overall liquid film coefficient) might perhaps be related to the fourth power of temperature on a bulk scale, similar to the finding by Professor Haslam on a laboratory scale, as described above;
- John Hunter [Hunter 1979] related $K_L a$ to viscosity via a turbulence index G;
- It was then thought that viscosity might be related to the fourth power temperature and a plot of the inverse of absolute viscosity against the fourth power of temperature up to near the boiling point of water gives a straight line;
- The interfacial area of bubbles per unit volume of bulk liquid under aeration is a function of the gas supply volumetric flow rate which is in turn a first-order function of temperature;
- It was then thought that $K_L a$ might be directly proportional to the 5th power of absolute temperature and indeed so, as verified by Hunter's data described in the following Section 2.4.1 (Fig. 2.2); the relationship, however, was not exact because the data plot deviates from a straight line at the lower temperature region;
- Adjustment of the initial equation based on observations of the behavior of certain other intensive properties of water in relation to temperature improved the linear correlation with a correlation coefficient of $R^2 = 0.9991$ (Fig. 2.3);
- The relationship is based on fixing (holding constant) all the extensive factors affecting the mass transfer mechanism. Specifically, $K_L a$ is dependent on the gas mass flow rate. However, since Hunter's data has slight variations in the gas mass flow rate over the temperature tests, normalization to a fixed gas flow rate improves the accuracy for the straight line passing through the origin with $R^2 = 0.9994$ (Fig. 2.4).

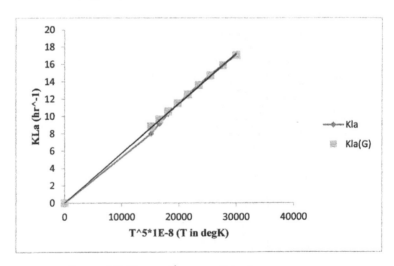

FIGURE 2.2 $K_L a$ vs. 5th power of absolute temperature

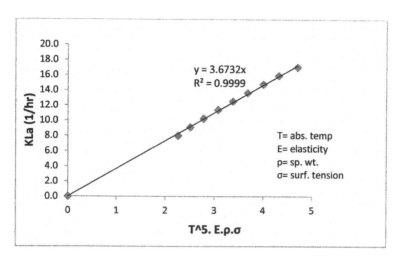

FIGURE 2.3 $K_L a$ vs. temperature, modulus of elasticity, density and surface tension

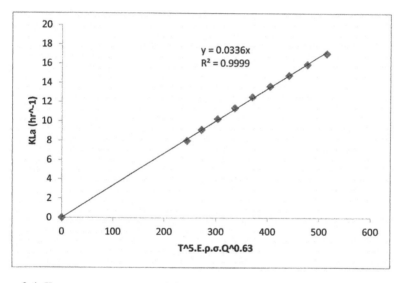

FIGURE 2.4 $K_L a$ vs. temperature, modulus of elasticity, density, surface tension, gas flow rate

Based on the above reasoning, data analysis as described in detail in the following sections confirmed the validity of (Eq. 2.1), but only for the special case where P_S is at or close to atmospheric pressure (i.e. $P_S = 1$ atm), assuming Hunter's tests were carried out at 1 atm. The experiments described in this paper have not proved that $K_L a$ is inversely related to P_S. The author advances a hypothesis that $K_L a$ is inversely proportional to equilibrium concentration (C_S), which can be related to pressure, which in turn is related to the depth of a column of water. Since saturation concentration is directly proportional to pressure (Henry's Law), $K_L a$ must be inversely proportional to pressure, if the reciprocity relationship between $K_L a$ and C_S is true. This is discussed in another paper published by the author [Lee 2018] and in the following chapters where relevant.

Furthermore, the concept of equilibrium pressure P_S and how to calculate P_S must be clarified for a bulk column of liquid. (The details for the pressure adjustment are given in ASCE 2-06 Section 8.1 and ANNEX G [ASCE 2007]). Insofar as the current temperature correction model has not accounted for any changes in P_S due to temperature, this manuscript has assumed that P_S is not a function of temperature for a fixed column height and therefore does not affect the application of (Eq. 2.1) for temperature correction.

2.2   THEORY

The Liquid Film Coefficient (k_l) can be related to the Overall Mass Transfer Coefficient (K_L) for a slightly soluble gas such as oxygen. For any gas-liquid interphase, Lewis and Whitman's two-film concept proved to be adequate to derive a relationship between the total flux across the interface and the concentration gradient, given by:

$$N_0 = K_L \times (C_s - C) \tag{2.9}$$

It can be proven mathematically that the bulk mass transfer coefficient is related to the respective film coefficients by the following equation:

$$K_L = \frac{k_g k_l}{H k_l + k_g} \tag{2.10}$$

where k_l and k_g are mass transfer coefficients for the respective films that correspond directly to their diffusivities and film thicknesses. H is the Henry's Law constant.

When the liquid film controls, such as for the case of oxygen transfer or other gas transfer that has low solubility in the liquid, the above equation is simplified to

$$K_L = k_l \tag{2.11}$$

This means that the gas transfer rate on a macro scale is the same as in a micro scale when the liquid film is controlling the rate of transfer due to the fact that the liquid film resistance is considerably greater than the gas film resistance. The four equations Prof Haslam developed are given below:

$$k_g = 290 \times MU^{0.8} T^{-1.4} \tag{2.12}$$

$$k_g = 0.72 \times MU^{0.8} \left(\frac{s}{\mu}\right)^{0.667} \tag{2.13}$$

$$k_l = 5.1 \times 10^{-7} \times T^4 \tag{2.14}$$

$$k_l = 37.5 \left(\frac{s}{\mu}\right)^{0.667} \tag{2.15}$$

Equations (2.12) and (2.13) are not important, since any changes in the rate of transfer in the gas film are insignificant compared to the changes in the liquid film for a slightly soluble gas such as oxygen. Equation (2.15) relates the liquid film to two physical properties of water, density (s) and viscosity (u). Equation (2.14) is most useful since it relates the mass transfer coefficient directly to temperature, irrespective of the gas flow velocity (U) or the molecular weight (M), and appears to be independent of Equation (2.15). Because the interphase concentrations are impossible to determine experimentally, only the overall mass transfer coefficient K_L can be observed in his apparatus. However, by substituting the values of the film coefficients calculated using the above equations into Equation (2.10), excellent agreement was found between the observed values of the overall coefficients and those calculated. Because of Equation (2.14), it can be concluded that the overall mass transfer coefficient in a bulk liquid is proportional to the fourth power of temperature, given by:

$$K_L = k' \times T^4 \tag{2.16}$$

where k' is a proportionality constant.

For spherical bubbles, the interfacial area (a) is given by:

$$a = \frac{\frac{Q_a}{\pi} \times 6}{d_b^3} \times \frac{\pi d_b^2}{V} \times t_c \tag{2.17}$$

where Q_a = average gas volumetric flow rate (m³/min); d_b = average diameter of bubble (m, mm); t_c = contact time of bubble with liquid; V = tank volume.

The contact time is dependent upon the path of the bubble through the liquid and can be expressed in terms of the average bubble velocity v_b and the liquid depth Z_d:

$$t_c = \frac{Z_d}{v_b} \tag{2.18}$$

where, v_b = average bubble velocity, (ms⁻¹)

The area of bubble interface per unit of tank volume V is then

$$a = 6 \times \frac{Q_a}{d_b v_b V} \times Z_d \tag{2.19}$$

This shows that for a given tank depth, and a fixed aeration system, 'a' is proportional to the gas flow rate Q_a. The mass transfer coefficient is dependent on the volumetric gas flow rate which changes with temperature and pressure-the higher the gas flow rate, the faster the transfer rate.

Average Volumetric Gas Flow Rate

The average gas flow rate is dependent on the test temperature of the bulk liquid. With this in mind, Q_a can be determined using the ASCE standard 2-06 [ASCE 2007] as follows:

Combining Eq. A.1b and Eq. A.2b in Section A.5.1 of Annex A where they were written as:

$$Q_1 = Q_P \left(\frac{T_1 P_P}{T_P P_1} \right) \tag{2.20}$$

$$Q_S = \frac{Q_1 T_S P_1}{T_1 P_S} \qquad (2.21)$$

where,

Q_S = gas flow rate given at standard conditions (i.e. the feed gas mass flow rate), (Nm^3/min)

Q_1 = gas flow at the gas supply system

Q_P = gas flow at the point of flow measurement (at the diffuser depth)

P_S = standard air pressure, 1.00 atm (101.3 kPa)

P_1 = ambient (gas supply inlet) atmospheric pressure

P_P = gas pressure at the point of flow measurement

T_S = standard air temperature (293 K for U.S. practice)

T_1 = ambient (gas supply inlet) temperature, K (=°C + 273)

T_P = gas temperature at the point of flow measurement

By substituting (Eq. 2.20) into (Eq. 2.21), we have

$$Q_P = Q_S \left(\frac{P_S}{P_P}\right)\left(\frac{T_P}{T_S}\right) \qquad (2.22)$$

Assuming the mass amount of gas is conserved, as the bubbles rise to the surface, Boyle's Law states that the volume is increased as the liquid pressure decreases, giving the following:

$$Q_{top} = \left(\frac{P_P}{P_b}\right) Q_S \left(\frac{P_S}{P_P}\right)\left(\frac{T_P}{T_S}\right) \qquad (2.23)$$

where P_b is the barometric pressure over the tank and Q_{top} is the volumetric flow rate at the top of the tank. The average gas flow rate over the entire column is therefore obtained by averaging of the gas flow rates given by Eq. (2.22) and Eq. (2.23) and is calculated by $Q_a = 1/2(Q_{top} + Q_P)$ and so,

$$Q_a = \frac{\frac{Q_S P_S T_P}{2}}{T_S} \times \left(\frac{1}{P_P} + \frac{1}{P_b}\right) \qquad (2.24)$$

Since $P_S = 1.01325 \times 10^5$ N/m^2 and $T_S = 293.15$ K (20°C),

therefore, substituting the standard values into (Eq. 2.24) yields the average gas flow rate in terms of the standard gas flow rate as:

$$Q_a = Q_S \times 172.82 \times T_P \left(\frac{1}{P_P} + \frac{1}{P_b}\right) \qquad (2.25)$$

Combining Eq. (2.16), Eq. (2.19) and Eq. (2.25) yields:

$$K_L a = k' T^4 \times 6 Q_S \times 172.82 \times T \times \left(\frac{1}{P_P} + \frac{1}{P_b} \right) \frac{Z_d}{d_b v_b V} \qquad (2.26)$$

Grouping all the numerical constants together into one single term, we have

$$K_L a = k'' Q_S \times T^5 \times \left(\frac{1}{P_P} + \frac{1}{P_b} \right) \frac{Z_d}{d_b v_b V} \qquad (2.27)$$

where k'' is another proportionality constant. This equation (Eq. 2.27) illustrates the 5th power temperature correction relationship as shown in (Eq. 2.1) for a fixed height Z_d, and volume V, assuming the pressures and the average bubble diameter (d_b) and velocity (v_b) do not change substantially over the temperature range tested.

As stated above, the response of $K_L a$ to temperature is affected by the behavior of the water properties that are the other variables that also affect the 5th order temperature relationship. As the temperature drops, the density of water (ρ) increases, and the maximum density is at about 4°C. Similarly, the surface tension (σ) also increases with the decrease of temperature. However, the modulus of elasticity (E) decreases as the temperature decreases. This is because the modulus of elasticity is proportional to the inverse of compressibility, which increases as the water approaches the solid state. Compressibility of water is at theminimum at around 50°C. Combining all the three variables in response to temperature with the 5th order relationship would result in a curve that resembles the error structure in Hunter's experiment as described in Section 2.4 below. These changes in water properties with respect to temperature are shown in Figs. 2.5, 2.6 and 2.7. The variability of the compound parameter $(E \rho \sigma)$ with temperature is also shown in Fig. 2.7 for the elasticity curve. Taking into account the changes in water properties in response to temperature, (Eq. 2.27) can be simplified to:

$$(K_L a)_T = K \times T^5 \times \frac{E \rho \sigma}{P_S} \qquad (2.28)$$

where the symbols are as defined in (Eq. 2.1). The inverse relationship between $(K_L a)_T$ and P_S is a hypothesis, based on the assumption that the mass transfer coefficient $(K_L a)_T$ and the solubility CST are inversely proportional to each other.

2.3 MATERIALS AND METHODS

To derive a temperature correction model, there are two ways. One is to use the solubility law derived from the solubility table for water (section 2.5), and the knowledge that $K_L a$ is inversely proportional to C_S, under a reasonable temperature boundary range. The other method is by use of examination and interpretation of actual data performed by numerous investigators, such as Hunter's data [Hunter 1979], on the relationship between $K_L a$ and temperature.

The new model for the correction number N as defined by (Eq. 2.3) is based on the 5th order proportionality. Numerous investigators have performed experiments of $K_L a$ determination at different test water temperature, ranging from 0°C to 55°C.

These data appear to support the hypothesis that $K_L a$ is proportional to the 5th power of absolute temperature for a range of temperatures close to 20°C and higher. For temperatures close to 0°C, however, the water properties begin to change in anticipation of a change of physical state (see Figs. 2.5, 2.6, 2.7 below).

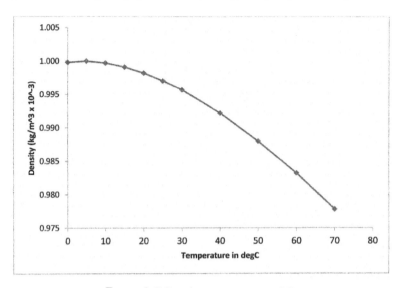

FIGURE 2.5 Density vs. temperature°C

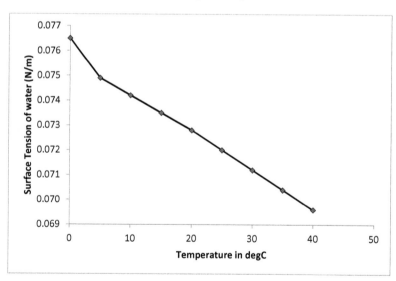

FIGURE 2.6 Surface tension vs. temperature°C

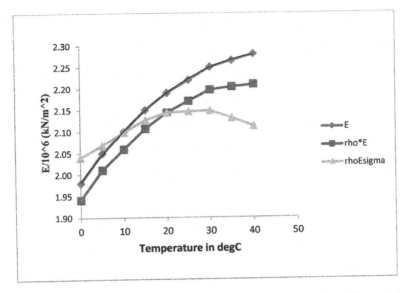

FIGURE 2.7 Modulus of elasticity vs. temperature°C (top curve) (Note: E is modulus of elasticity; rho is density of water; sigma is surface tension)

This change from a liquid state to a solid state at this low temperature is unique to water. However, by incorporating these changes of the relevant properties into the K_La equation, as described previously, it becomes possible to find a high degree of correlation for the data interpretation.

The following paragraphs describe the derivation method to arrive at the proposed temperature correction model by use of experimental data. This derivation is purely based on data interpretation and data analysis using linear graphical verification, and is not derived theoretically.

2.3.1 Hunter's Experiment

Hunter [1979] performed an experiment for the case of laboratory-scale submerged turbine aeration systems. He derived an equation that relates K_La to the various extensive properties of the system and to viscosity, and correlated his data for a temperature range of 0–40°C. His method is described in the paper cited in the manuscript and in his dissertation: Hunter [1977].

2.3.2 Vogelaar et al.'s Experiment

The experiments performed by Vogelaar et al. [2000] consist of determining K_La using tap water for a temperature range of 20°C – 55°C using a cylindrical bubble column with an effective volume of 3 liters and subject to aeration flow rates of 0.15, 0.3, 0.45, and 0.56 vvm (volume air volume liquid^{-1} min^{-1}). The results for **one particular volumetric air flow rate (0.3 vvm)** among all the data are given in Table 2.3 further below.

The following section describes how the data from these two research workers have been used to develop the temperature correction equation for determining $(K_La)_{20}$ for any clean water test carried out in accordance with ASCE 2-06, and it is proposed that this new equation is to be used to replace the current equation as stated in ASCE 2-06 Section 8.1 and the relevant sections concerning the use of (θ) in the calculation of this important parameter $(K_La)_{20}$ – the standardized K_La at standard conditions as defined in the ASCE Standard.

2.4   RESULTS AND DISCUSSION

Hunter [1979] has suggested that turbulence can be related to viscosity as well as the aeration intensity that created the turbulence. In surface aeration, aeration intensity can be the power input to the water being aerated, while in subsurface diffused aeration, it is likely to be the air bubbles' flow rate. Therefore, for certain fixed power intensity, Hunter surmised that K_La is only a function of viscosity which in turn is a function of temperature. He created a mathematical model that related $(K_La)_T$ to viscosity at different temperatures from 0°C to 40°C. His results are given below in Table 2.1, where $K_La(G)$ are his modelled results. The model he used was expressed as:

$$K_La(G) = \left(4.04 + 0.00255G^2 \times \left(\frac{D}{T}\right)^4\right)Q^{.63} \tag{2.29}$$

where D/T is a geometric function. [Note that T in his equation is NOT temperature] $G^2 = P/V/\mu$ where μ is viscosity, and P is the power level (total power input into the water being aerated in ergs/s, and V is the volume of tank in cm^3). The term G was defined as the turbulence index. However, just as in solubility, it is erroneous to consider G as a function of viscosity because viscosity is an intensive property, not extensive. Changing the viscosity would not increase turbulence, in the same way turbulence does not affect viscosity for a fixed temperature. However, in his paper's attachment, he has theoretically derived a relationship between r, the rate of gas-liquid interfacial surface renewal, and the turbulence index G, that they are equal. Since K_L, the liquid film coefficient, is related to r, it can be concluded that turbulence affects the mass transfer coefficient, but this is not due to the apparent correlation between G and μ.

In this table, the observed K_La results are given in column 5. His modelled results are given in column 6. As one can see, his predicted results match up quite well with the true results for those tests carried out at 20°C and above. At the lower temperature range, however, his errors increase progressively as the temperature drops to the water freezing point. His results can be seen from the following plot in Figure 2.8 below:

TABLE 2.1 Hunter's experimental data (*Note: The air flow rate Q is back calculated from Eq. 2.29 for $D/T = 0.35$ and $P/V = 2000$)

$T(°C)$	Viscosity (Poise)	$T(K)$	$(T/1000)^5 \times 10^4$	$K_L a\,(h^{-1})$	$K_L a\,(G)\,(h^{-1})$	$Q\,(SCFH)^*$
		0	0	0	0	
0	0.01787	273.15	15.21	7.99	8.8	1.093
5	0.01519	278.15	16.65	9.12	9.64	1.1
10	0.01307	283.15	18.2	10.26	10.55	1.107
15	0.01139	288.15	19.87	11.39	11.51	1.113
20	0.01002	293.15	21.65	12.53	12.53	1.118
25	0.008904	298.15	23.56	13.66	13.6	1.124
30	0.007975	303.15	25.6	14.79	14.71	1.128
35	0.007194	308.15	27.79	15.93	15.87	1.132
40	0.006529	313.15	30.11	17.06	17.08	1.136

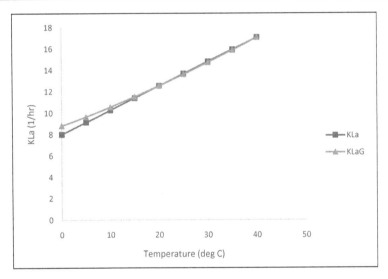

FIGURE 2.8 Hunter's data of $K_L a$ plotted vs. temperature°C

   Hunter did not explain why the errors in terms of percent difference become more pronounced toward the lower end of the temperature spectrum, since the turbulence index G has already accounted for the increase of viscosity due to temperature, and so if turbulence was only a function of viscosity, the changes due to viscosity to the mass transfer coefficient should have been taken care of in his equation. However, in his attachment, he did derive an equation that relates $K_L a$ not only to G, but also to other system variables which he had not defined. (Note: Hunter's formula did include the extensive properties as system variables in his experiment: geometry, power level, volume, gas flow rate. But while the extensive properties are important factors affecting $K_L a$, it is found in this study that the relationship between $K_L a$ and the intensive properties is always linear, and this linear relationship is independent of the extensive properties. The intensive properties are all temperature dependent). Hunter did not know of the 5th power model. Had he plotted his $K_La(G)$ values against the 5th power of absolute temperature, he would have been astonished to see a perfect straight line as shown in Figure 2.2 before.

Hunter's model is in fact correct if all the other system variables were fixed, so that K_La is only a function of viscosity. The other system variables are in fact the other properties of water, such as density, modulus of elasticity, and surface tension. As the liquid approaches its freezing point, it is subject to all the changes in these properties in precedence to the anticipated changes of physical states.

These changes in water properties can be seen by plotting the handbook values for these parameters at different temperatures, as shown in Figs. 2.5, 2.6 and 2.7.

These changes in the other properties of water explain why his data starts to deviate from a straight line when the temperature drops below 20°C. Figure 2.7 above shows what happens when Hunter's data is successively corrected for these changes. The final curve showing the product $\rho \cdot E \cdot \sigma$ (rhoEsigma) vs. T represents the correction by the product of density, elasticity, as well as surface tension. The curve resembles the error structure in Hunter's data (comparing col.5 and col.6 in Table 2.1). So, when the mass transfer coefficient data is plotted against T^5 multiplied by the correction factor F, which in this case is given by $\rho \cdot E \cdot \sigma$, a much better linear relationship is obtained, as is shown in Figure 2.3 before.

2.5 CALCULATION

2.5.1 Hunter's Data

Data0 analysis based on Hunter's experiments [Hunter 1979] has supported that K_La (the oxygen mass transfer coefficient) needs to be corrected for surface tension in addition to E and ρ. The effect of surface tension on K_La is more pronounced towards the lower temperature region (below 20°C and as it gets closer to the melting point (freezing point) of the solvent (Fig. 2.6) where surface tension increases rapidly as the temperature decreases). From (Eq. 2.8), it can be seen that $(K_La)_{20}$ can be calculated based on a single test data on K_La. It is important to note that the temperature correction factor N should not be calculated as the ratio μ_{20}/μ_T, but as $(K_La)_{20}/(K_La)_T$; therefore, at 20°C

$$(K_La)_{20} = K'' \frac{(T.E.\rho\sigma)_{20}}{P_{S_{20}}} \times \left(\frac{1}{\mu_{20}} + 0.7815 \right) \qquad (2.30)$$

By eliminating K'' and assuming $P_S = P_{S_{20}}$, therefore,

$$(K_La)_{20} = K_La \cdot \frac{[E\rho\sigma \times T \times fn(u)]_{20}}{[E\rho\sigma \times T \times fn(u)]_T} \qquad (2.31)$$

or

$$N = \frac{[E\rho\sigma \times T \times fn(u)]_{20}}{[E\rho\sigma \times T \times fn(u)]_T} \qquad (2.32)$$

where $fn(u)$ is given by $(1/u + 0.7815)$

Similarly, (Eq. 2.1) for the 5th power model can be used to calculate $(K_L a)_{20}$ and result in the following Table 2.2 and the following equation (Eq. 2.33):

$$(K_L a)_{20} = K_L a \times \frac{[E\rho\sigma T^5]_{20}}{[E\rho\sigma T^5]_T} \tag{2.33}$$

$$N = \frac{[E\rho\sigma T^5]_{20}}{[E\rho\sigma T^5]_T} \tag{2.34}$$

The correction number values are given in column 10 in Table 2.2. It should be noted that even without including these additional variables E, ρ, σ, the 5th power model already gives a very good fit to the experimental data. In fact, the fifth power model gives a slightly better fit than the enhanced model for temperatures above 20°C. The effects of these other physical properties seem to wane as the temperature increases toward the boiling point region. This is apparent from Hunter's model as shown in Table 2.1 (comparing column 5 and column 6), where the prediction error of his model becomes negligible when compared with the observed data when temperature is above 20°C. The enhanced model plot that is inclusive of the factors E, ρ, σ, is given in Figure 2.3 with a high degree of correlation ($R^2 = 0.9991$).

TABLE 2.2 Simulated results for the prediction of $(K_L a)_{20}$ by the 5th power model

1	2	3	4	5	6	7	8	9	10	11
T	T	$T^6/10^{12}$	$K_L a$	ρ	$E/10^6$	σ	$E \cdot \rho \cdot \sigma$	*Corr. No. F	*Corr. No. N	$(K_L a)_{20}$
[degC]	[K]	*	[1/hr]	[kg/m³]	[kN/m²]	[N/m]				
0	273.15	1.5206	7.99	999.8	1.98	0.0765	151.44	1.051	1.496	11.95
5	278.15	1.6649	9.12	1000.0	2.05	0.0749	153.55	1.036	1.348	12.29
10	283.15	1.8200	10.26	999.7	2.10	0.0742	155.77	1.022	1.215	12.47
15	288.15	1.9865	11.39	999.1	2.15	0.0735	157.88	1.008	1.099	12.51
20	293.15	2.1650	12.53	998.2	2.19	0.0728	159.15	1.000	1.000	12.53
25	298.15	2.3560	13.66	997.0	2.22	0.0720	159.36	0.999	0.918	12.54
30	303.15	2.5603	14.79	995.7	2.25	0.0712	159.51	0.998	0.844	12.48
35	308.15	2.7785	15.93	993.9	2.27	0.0704	158.48	1.004	0.782	12.46
40	313.15	3.0114	17.06	992.2	2.28	0.0696	157.45	1.011	0.727	12.40

*Note: $F = (E\rho\sigma)_{20}/(E\rho\sigma)_T$
$N = F \cdot (T_{20}/T)^5$ or $(K_L a)_{20} = K_L a \cdot N$
$T^5 = (T/1000)^5 \times 1000$

Using the predicted $(K_L a)_{20}$ based on the 5th power model and plotting the simulated results with test temperatures, the following Figure 2.9 is obtained:

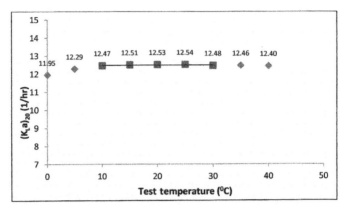

FIGURE 2.9 Simulated results for the prediction of $(K_La)_{20}$ [Hunter 1979]

This shows that the variations in the prediction of $(K_La)_{20}$ based on the various tests at different temperatures are very small and in fact are much smaller than would be obtained from using the current ASCE model. Figure 2.10 and Figure 2.11 below show the discrepancies between the various models ($\theta = 1.024$, $\theta = 1.018$, and the 5th power model) even further.

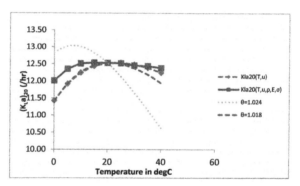

FIGURE 2.10 Comparison of $(K_La)_{20}$ as predicted (0°C~40°C) by various models [Hunter 1979]

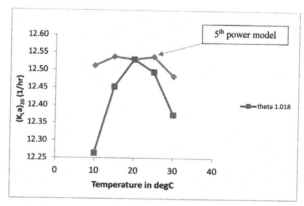

FIGURE 2.11 Comparison of $(K_La)_{20}$ predicted by two close models between 10°C and 30°C

It should be clear from these graphs that the 5th power model is superior to the current model that uses the theta (θ) correction factor, for temperatures between 10°C and 30°C, which is the temperature range stipulated in ASCE 2-06.

The plot in Fig. 2.3 showing the linear relationship between the mass transfer coefficient and the 5th power temperature function can be further improved if the $K_L a$ data are normalized to the same gas flow rate (data given in Table 2.1 col. 7 for the flow rates). Hunter's equation has stipulated that the predicted $K_L a (G)$ is proportional to the value of $Q^{0.63}$ and so plotting $K_L a$ against the function $T^5 \cdot E \cdot p \cdot \sigma$ together with $Q^{0.63}$ further improves the correlation as was shown in Figure 2.4. Therefore, based on Hunter's experiment, and the good correlation results as shown in Figure 2.3 ($R^2 = 0.9991$) and Figure 2.4 ($R^2 = 0.9994$), it can be concluded that for a fixed mass gas flow rate, the mass transfer coefficient under different test temperatures can be calculated by (Eq. 2.1). Therefore, the correction number N can be calculated by simple proportion as given by (Eq. 2.34). This equation has assumed that P_S remains constant at different temperatures.

2.5.2 Vogelaar's Data

Similarly, Vogelaar's Experiment [Vogelaar et al. 2000] showed excellent correlation between $K_L a$ and C_S for temperatures above 20°C, and Vogelaar's experimental result is given in Table 2.3 below:

TABLE 2.3 Vogelaar's experimental results

$T(°C)$	$T(K)$	$(T/1000)^5 \times 10^4$	C_S(mg/L)	$1/C_S$	$K_L a\,(h^{-1})$
	0	0		0	0
20	293.15	21.65	9.19	0.1088	22.4+/−0.4
30	303.15	25.60	7.43	0.1346	26.0+/−0.1
40	313.15	30.11	6.5	0.1538	30.6+/−0.2
55	328.15	38.05	5.15	0.1942	38.8+/−1.5

Figure 2.12 below shows a plot of $K_L a$ vs. $T^5 \cdot (E\rho\sigma)$ and the correlation is excellent with $R^2 = 0.9975$, assuming $P_S = 1$ atm. However, it is not as good as Hunter's data using the same model. At 55°C, the deviation from the straight line is larger than the other data points. It is not clear why this is so. It could be that the distribution of the experimental errors is not even, or that the gas flow rate is not quite identical at this point. In any case, the prediction of $(K_L a)_{20}$ is still much better than using $\theta = 1.024$ or any other values except 1.016, as shown in Fig. 2.13.

At 55°C, the discrepancy between the theta (θ) model and the 5th power model is greater than 30%. When plotting the predicted $(K_L a)_{20}$ values using the various models (5th power, $\theta = 1.024$, $\theta = 1.018$, $\theta = 1.016$), the following graph is obtained (Fig. 2.13). As seen from this plot, the 5th power model predicts a series of consistent values of $(K_L a)_{20}$, whereas the (θ) model using $\theta = 1.024$ gives very poor results. Although using $\theta = 1.018$ improves the prediction, it is still not as good as the 5th

power model. The difficulty of using the (θ) model is that the value of θ must be pre-determined by testing which is the major disadvantage of this model.

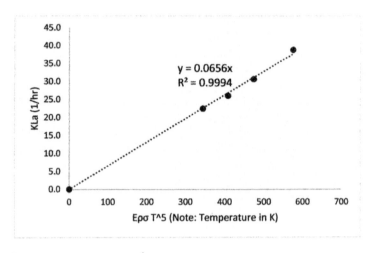

FIGURE 2.12 $K_L a$ against 5th power of temperature [Vogelaar et al. 2000]

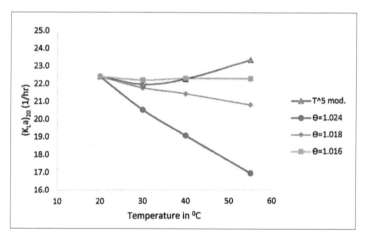

FIGURE 2.13 Comparison of simulated results for $(K_L a)_{20}$ [Vogelaar et al. 2000]

2.5.3 Methodology for Temperature Correction

A new model to improve the temperature correction for $K_L a$ used in ASCE Standard 02 has been developed. Based on data analyses of two researchers' work, it can be seen that the new model gives excellent simulated results for $(K_L a)_{20}$ based on series of tests at increasing water temperatures, compared to the other models using the same data as seen in Fig. 2.10 and Fig. 2.13. The major function of this model is to predict $K_L a$ for any changes in temperature so that $(K_L a)_{20}$ can be predicted from any one single test at a specific temperature, and therefore would replace the current model in ASCE 2-06 with a higher degree of accuracy. For a certain equilibrium level (d_e) where the equilibrium pressure is at (P_S), the model is expressed by (Eq. 2.1), in which the proportionality constant K is dependent on the extensive properties of the aeration system, such as gas flow rate, bubble size and other characteristics of the system. This equation is not complete because temperature also affects the volumetric gas flowrate Q_a. Therefore, as a result of the foregoing analysis, the formula for estimating $(K_L a)_{20}$ based on any given test at a test temperature T °C is given by:

$$\langle K_L a \rangle_{20} = (K_L a)_T \left(\frac{(\rho E \sigma)_{20}}{(\rho E \sigma)_T} \right) \left(\frac{T_{20} + 273}{(T_T + 273)} \right)^5 \left(\frac{P_{ST}}{P_{S20}} \right) \left(\frac{Qa_{20}}{Qa_T} \right) \quad (2.35)$$

where T is expressed in °C. For a series of tests under the same barometric pressure, as in Hunter's experiment, the change in (P_S) due to temperature is likely to be small, and so the ratio $P_{S_T}/P_{S_{20}}$ can be cancelled. Temperature affects the gas volumetric flow rate, even for a fixed mass gas flow rate. But for a fixed gas flow rate and fixed pressure, and since the values of E, ρ and σ for water are fixed, a table of correction factors can be compiled to make the application easy, as shown in Table 2.4 below:

TABLE 2.4 Table of correction factors for the temperature correction model (F, N)

1	2	3	4	5	6	7	8
T(°C)	T(K)	ρ (kg/m³)	$E/10^6$ (kN/m²)	σ (N/m)	$E \cdot \rho \cdot \sigma$	F	$N = F \cdot \left(\frac{T_{20}}{T} \right)^5$
0	273.15	999.8	1.98	0.0765	151.44	1.051	1.496
5	278.15	1000.0	2.05	0.0749	153.55	1.036	1.348
10	283.15	999.7	2.10	0.0742	155.77	1.022	1.215
15	288.15	999.1	2.15	0.0735	157.88	1.008	1.099
20	293.15	998.2	2.19	0.0728	159.15	1.000	1.000
25	298.15	997.0	2.22	0.0720	159.36	0.999	0.918
30	303.15	995.7	2.25	0.0712	159.51	0.998	0.844
40	313.15	992.2	2.28	0.0696	157.45	1.011	0.727
50	323.15	988.0	2.29	0.0679	153.63	1.036	0.636
60	333.15	983.2	2.28	0.0662	148.40	1.072	0.566

Although in reality, $(K_L a)_{20}$ should be normalized to the same average gas volumetric flow rate in order to be more precise, in as much as the current equation for correcting $K_L a$ in ASCE 2-06 has not accounted for changes in gas flow rate nor any other effects such as (P_S), it is recommended that, for the time being, it is sufficiently accurate to replace the current equation by a simplified equation, in the effort to standardize the mass transfer coefficient to a standard condition of 20°C and standard atmospheric pressure. For a fixed mass gas flow rate, the equation becomes:

$$(K_L a)_{20} = (K_L a)_T \left(\frac{(\rho E \sigma)_{20}}{(\rho E \sigma)_T} \right) \left(\frac{T_{20} + 273}{(T_T + 273)} \right)^5 \tag{2.36}$$

where T is again expressed in terms of degree Celsius. Equation (2.36) is the proposed model for temperature correction for $K_L a$ to be used on ASCE Standard 02, where $F = (E \rho \sigma)_{20}/(E \rho \sigma)_T$ and the correction number N would be given by $N = F \cdot (T_{20}/T)^5$, where in the application of the temperature correction model for $(K_L a)_{20}$, $(K_L a)_{20}$ is obtained by multiplying $(K_L a)_T$ by the correction number N in column 8 at the test temperature T. The correction factors can be plotted against temperature for easy use, such as in Fig. 2.14 below.

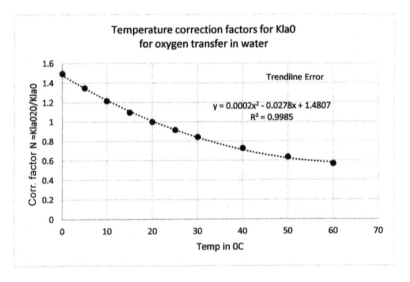

FIGURE 2.14 Temperature correction model in graphical form
(**Note:** for the meaning of $K_L a_0$, refer to Chapter 3)

2.6 THE SOLUBILITY MODEL

As mentioned in the introduction, the author advocates the hypothesis that solubility is inversely proportional to $K_L a$. The foregoing sections have established that $K_L a$ is related directly to the 5th order of temperature. If this hypothesis is true, then one would expect the solubility also bears a 5th order relationship with temperature, but in an inverse manner. The following sections illustrate that solubility is indeed related to the 5th order of temperature using published scientific data. This section is significant for 2 reasons:

First, a new physical law is discovered. According to the Oxford English dictionary, a physical law "is a theoretical principle deduced from particular facts, applicable to a defined group or class of phenomena, and expressible by the statement that a particular phenomenon always occurs if certain conditions be present." The rationale behind the solubility law is similar to the Universal Gas Law which is in fact an extension of Boyle's Law or Charles' Law. Just as Boyle's Law states that the absolute pressure exerted by a given mass of an ideal gas is inversely proportional to the volume it occupies if the temperature and amount of gas remain unchanged within a closed system, so the Universal Gas Law states that, for a fixed mass of gas, volume is inversely proportional to pressure, and directly proportional to temperature.

The solubility law relates oxygen solubility in water to the 5th power of temperature, and also to certain properties of water (Eq. 2.37 below). This relationship has not appeared in any literature until now and it is therefore accurate to claim that the 5th power inverse relationship was up to now unknown prior to this manuscript. The author believes that the reason this solubility law was not discovered earlier like the gas law is that in the gas law, all the parameters are first order and can easily be verified experimentally. In the solubility law, the inverse 5th power phenomenon is not directly observable. Furthermore, the solubility law deals with the interaction of two phases and two species—solute gas and solvent liquid, whereas the gas law deals with only a single gas phase. It is one thing to test a model once it has been discovered, but quite another to find the physical law in the first place.

Second, the law can be applied to real situations in wastewater treatment, and in many bioreactor processes. One of the major applications is the prediction of oxygen transfer in water. The topic discussed in this manuscript is about gas transfer in water, how much and how fast, in response to changes in water temperature. The hypothesis is that K_La and C_S are in fact inversely proportional to each other. This paper demonstrates how the discovered physical law for gas solubility can be compared with the temperature correction model for K_La based on experimental data [Hunter 1979] [Vogelaar et al. 2000] that will prove the hypothesis that K_La is inversely and linearly proportional to C_S.

2.7 DESCRIPTION OF THE OXYGEN SOLUBILITY MODEL

Oxygen solubility in water is affected by both temperature and pressure. The influence of temperature on the solubility of gases is predictable. The Benson and Krause (1984) oxygen solubility model is well known and is adopted by the USGS (United States Geological Survey) and ASCE 2-06. This model, however, is only applicable for a special case where the atmospheric pressure is at the standard pressure of 101.3 kN/m^2. The model is empirical and based on data collected for that pressure only.

Apart from temperature, pressure has a strong effect on the solubility of a gas. For a fixed temperature, the relationship between solubility and pressure is governed by Henry's Law. Hitherto, however, equation has not existed that combines both effects into one single formula.

Henry's Law, which states that the solubility (or saturation concentration) of a gas in a liquid is directly proportional to the partial pressure of the gas if the temperature is constant, can be explained by Le Chatelier's principle in a body of water. The principle states that when a system at equilibrium is placed under stress, the equilibrium shifts to relieve the stress. In the case of saturated solution of a gas in a liquid, equilibrium exists whereby gas molecules enter and leave the solution at the same rate. When the system is stressed by increasing the pressure of the gas, more gas molecules go into solution to relieve that increase. This happens at the lower regions of a body of water such as an aeration tank or a lake well-mixed. Conversely, when the pressure of the gas is decreased, more gas molecules come out of solution to relieve the decrease and this happens at the upper regions.

The solubility law proposed herewith is an extension of Henry's Law. The proposed solubility law states that for any temperature and pressure, solubility is directly proportional to pressure, and inversely proportional to the fifth power of temperature in absolute temperature in Kelvin, and inversely proportional to density and modulus of elasticity of the solution, expressed as:

$$C_S = K \times \frac{P_S}{T^5 E \rho} \tag{2.37}$$

where,

K = proportionality constant. For water of zero salinity, K has a value of approximately $43.4 \text{ kg}^{-2} \cdot \text{N} \cdot \text{deg K} \cdot \text{m}^{-8} \cdot \text{atm}^{-1}$ when the units of the parameters are defined as C_S = mg/L; P_S = atm; $T = K(\text{Kelvin}) \times 10^{-3}$; $E = (\text{kN/m}^2) \times 10^{-6}$; ρ = kg/m^3.

Justification of this model, its derivation and verification, and the evaluation of the importance of temperature-dependent properties of water including the bulk modulus of elasticity of water and the density of water in the relationship will be presented later in Section 2.7 below. (See also Section 5.7 in Chapter 5 for the justification for using the 5th power model for correcting $K_L a$). Since this solubility law is newly discovered in the scientific community, it should be given a name such as the Law of Oxygen Solubility in Pure Water, and the constant K should be called the Oxygen-Water Solubility Law Constant.

2.8 ANALYSIS

There are many ways to confirm a physical law, once it has been discovered. For example, thermodynamic data could be used such as enthalpy and entropy, and the Gibbs free energy may be sufficient to verify the model. This method was used by Tromans [1998] in his derivation of his model of oxygen solubility in pure water. In this manuscript, graphical methods using linear proportions are used. Solubility data for other gases and other liquids are available, so that it should be possible to test the law on other media in order to determine whether the law can be applied to some other liquids. For example, oxygen solubilities in water of different chlorinities (salinities) are given by USGS, as well as in ASCE 2-06.

In the Office of Water Quality Technical Memorandum 2011.03 [Rounds 2011], it was announced that the equations that had traditionally been used by the U.S. Geological Survey (USGS) to predict the solubility of dissolved oxygen (DO) in water result in slight discrepancies between values predicted for DO solubility by USGS tables and computer programs compared with values computed by following the methods listed in Standard Methods for the Examination of Water and Wastewater (American Public Health Association 2005) (Standard Methods). Subsequent analysis resulted in a well-documented recommendation to replace the Weiss (1970) equations with the equations developed by Benson and Krause (1984).

The Benson and Krause (1984) oxygen-solubility formulations (now adopted by USGS) are documented in equations 1 and 7 through 11 of the Attachment to the Technical Memorandum. The equations adopted by the USGS and now in line with the Standard Methods are summarized in the following form:

$DO = DO_0 x F_S x F_P$, where the dissolved oxygen (DO) concentration in mg/L is represented as a baseline concentration in freshwater (DO_0) multiplied by a salinity correction factor (F_S) and a pressure correction factor (F_P). All three terms are a function of water temperature. In addition, the salinity correction factor is a function of salinity and the pressure correction factor is a function of barometric pressure. For freshwater (salinity = 0%) and standard pressure (1 atm), the salinity and pressure factors are equal to 1.0.

$$DO_0 = \exp\left(-139.3441 + \frac{1.575701 \times 10^5}{T} - \frac{6.642308 \times 10^7}{T^2} + \frac{1.243800 \times 10^{10}}{T^3} - \frac{8.621949 \times 10^{11}}{T^4}\right)$$

(2.38)

The salinity correction factor and the pressure correction factor are given by:

$$F_S = \exp\left[-S \cdot \left(0.017674 - \frac{10.754}{T} + \frac{2140.7}{T^2}\right)\right]$$

(2.39)

$$F_P = \frac{P - P_{vt}}{1 - P_{vt}} \times \frac{1 - \theta_0 \cdot P}{1 - \theta_0}$$

(2.40)

where S is salinity in parts per thousand (%) and T is temperature in Kelvin. P is the barometric pressure in atmospheres, P_{vt} is the vapor pressure of water in atmospheres, and θ_0 is related to the second virial coefficient of oxygen. Using the above equations, it was possible to construct a solubility table similar to the published Table CG-1 as given in ASCE 2-06. Such a constructed table for zero salinity is given in Table 2.5 col. 2 [Metcalf and Eddy, 2nd Edition] below:

TABLE 2.5 Physical properties of water at various temperatures

T (°C)	$C_{S(T)}$ (mg/L)	ρ (kg/m³)	$E/10^6$ (kN/m²)	σ (N/m)	$\mu \times 10^3$ (N·s/m²)	$\gamma \times 10^6$ (m²/s)	P_v (kN/m²)
0	14.62	999.8	1.98	0.0765	1.787	1.785	0.61
5	12.77	1000	2.05	0.0749	1.518	1.519	0.87
10	11.29	999.7	2.10	0.0742	1.307	1.306	1.23
15	10.08	999.1	2.15	0.0735	1.139	1.139	1.70
20	9.09	998.2	2.19	0.0728	1.002	1.003	2.34
25	8.26	997.0	2.22	0.0720	0.890	0.893	3.17
30	7.56	995.7	2.25	0.0712	0.798	0.800	4.24
40	6.41	992.2	2.28	0.0696	0.653	0.658	7.38
50	5.49	988.0	2.29	0.0679	0.547	0.553	12.33
60	4.71	983.2	2.28	0.0662	0.466	0.474	19.92

The other data pertaining to the physical properties of water, as shown in Table 2.5, is from the standard handbook and textbook [ASCE 2007] [Benson and Krause 1984], which enabled calculating:

- a temperature correction function: $T^5 \cdot E \cdot \rho / P_S$, and
- the reciprocal of solubility, or insolubility

When the insolubility $(1/C_S)$ is plotted against the temperature function at P_S = 1 atm, a straight line passing through the origin is obtained with the correlation $R^2 = 0.9998$ (graph not shown). The significance of this plot is that the extension of the linear plot passes through the point of origin at zero K. This does not mean the absolute temperature could reach the zero point (the molecular structure will have changed long before that), but such linear relationship offers a simple means of calculating solubility at any physical parameter of the solvent, by simple ratios. Since water changes from a liquid state to a solid state as the temperature approaches its melting point (freezing point), once the temperature drops past the melting point (normally 0°C at standard pressure), the law no longer holds and any projection past the solid state is therefore purely hypothetical. If the data of solubility is plotted against the inverse of the temperature correction function affecting solubility, the straight-line linear plot would be as shown in Figure 2.15 below. Therefore, the solubility law can be expressed either by the equation derived from plotting the insolubility, or expressed by the equation from plotting the data as in Figure 2.15. In the former method, the equation gives the insolubility of oxygen expressed by:

$$\frac{1}{C_S} = 0.02302 \cdot T^5 \times E \times \frac{\rho}{P_S} \tag{2.41}$$

where T is in K to the power 10^{-3}.

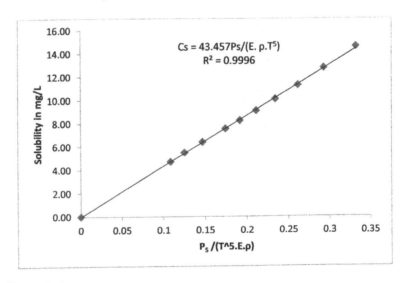

FIGURE 2.15 Solubility plot for water dissolving oxygen at P_S = 1 atm (1.013 bar)

In the latter case, the equation gives the solubility directly and is expressed by:

$$C_S = 43.457 \times \frac{P_S}{(T^5 E \rho)} \qquad (2.42)$$

TABLE 2.6 Solubility of oxygen in fresh water (salinity ~ 0) at different pressures and temperatures [Engineering ToolBox 2014]

T (deg C)	T (K)	1 bar	2 bar	4 bar
0	273.15	14.6	29.2	58.4
5	278.15	12.8	25.5	51.1
10	283.15	11.3	22.6	45.1
15	288.15	10.1	20.2	40.3
20	293.15	9.1	18.2	36.4
25	298.15	8.3	16.5	33.1
30	303.15	7.6	15.2	30.3
35	308.15	7.0	14.0	27.9
40	313.15	6.5	12.9	25.9
45	318.15	6.0	12.0	24.0
50	323.15	5.6	11.3	22.7

Henry's Law is applicable only to ideal solutions [Andrade 2013], and for an imperfect liquid subject to state changes at extreme temperatures, it is only approximate and limited to gases of slight solubility in a dilute aqueous solution with any other dissolved solute concentrations not more than 1 percent. At different pressures, the solubility will increase as shown in Table 2.6 [Engineering ToolBox 2014]. Similar plots (see Figure 2.16) for solubility at different pressures can be made using the data from the same source [Engineering Toolbox 2016]. From Fig. 2.16, the solubility is inversely proportional to the temperature function expressed in terms of T, E and ρ, so that

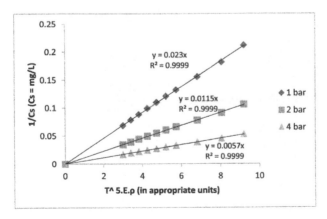

FIGURE 2.16 Comparison of oxygen solubility plots for various pressures

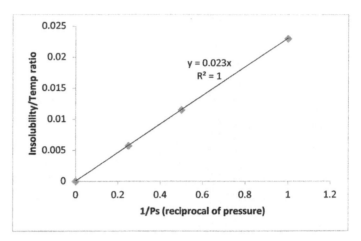

FIGURE 2.17 Proportionality constant K plotted against reciprocal of pressure

$$\frac{1}{C_S} = K(T^5 \cdot E \cdot \rho) \tag{2.43}$$

where K is a constant.

Plotting the K values (0.023 at 1 bar; 0.0115 at 2 bar; 0.0057 at 4 bar) from Figure 2.16 against the reciprocals of pressures, the following graph shown in Figure 2.17 is obtained.

Therefore,

$$K = 0.023\left(\frac{1}{P_S}\right) \tag{2.44}$$

Combining (Eq. 2.43) and (Eq. 2.44) we have

$$\frac{1}{C_S} = 0.023\left(\frac{1}{P_S}\right) \cdot (T^5 \cdot E \cdot \rho) \tag{2.45}$$

or

$$C_S = 43.478\frac{P_S}{T^5 \cdot E \cdot \rho} \tag{2.46}$$

(Eq. 2.46) is equivalent to (Eq. 2.42) showing that solubility is indeed proportional to pressure, in accordance with Henry's Law. The slight discrepancy in the K value arises from the two different sources of data, one from Bensen and Krause (1984), and the other from Engineering ToolBox (2001) [online], available at: https://www.engineeringtoolbox.com. But it is likely that the former is more accurate since the solubility data has two decimal places.

2.9   CONCLUSIONS

Based on the afore-mentioned literature review, the following conclusions are obtained:

- The primary intent of this manuscript is to replace the geometric technique as used in ASCE 2-06[ASCE 2007]. The current method that uses an assigned theta (θ) value for correcting the effects of temperature on oxygen transfer coefficient $(K_La)_{20}$ is empirical and attempts to lump all possible factors, such as changes in viscosity, surface tension, diffusivity of oxygen, geometry, rotating speed, type of aerators, etc. This empirical approach has produced a great variety of correction factors for theta. Therefore, a wide range of temperature correction factors is reported in the literature which has ranged from 1.008 to 1.047.ASCE 2-06 Commentary CG-3 recommends θ to be 1.024 and clean water testing should be at temperatures close to 20°C. When a value different from 1.024 is proposed, it usually requires justification by an extensive array of testing [Lee 1978] [Boogerd et al. 1990], and preferably testing in full-scale for a range of test temperatures as required, for the same conditions of each test. This may not be possible at all.
- The 5th power model developed is mechanistic in nature. Unlike the conventional empirical model, it does not require the selection of an uncertain parameter (a priori) value, such as theta (θ). The correction number N is independent of the extensive properties of an aeration system in the estimation of $(K_La)_{20}$, whereas the correction number for the θ model cannot be applied universally and pertains to the system that was used to obtain the parameter only. The new model should prove to be valid for other similar testing, especially in full-scale, because the resultant $(K_La)_{20}$ is dependent only on temperature and the other intensive properties of the fluid, if the extensive properties are fixed.
- For the temperature correction model, a formula is derived as defined by (Eq. 2.36):

$$(K_La)_{20} = (K_La)_T \left(\frac{(\rho E \sigma)_{20}}{(\rho E \sigma)_T} \right) \left(\frac{T_{20} + 273}{(T_T + 273)} \right)^5 \qquad (2.47)$$

The improvement of this model relative to the old model as given by (Eq. 2.2) is readily apparent when plotting the simulated $(K_La)_{20}$ for both models on thesame plot, as shown in Figs. 2.8, 2.9, 2.10 and 2.13. The prediction error is within 1% for the temperature range between 10°C and 55°C. This is assuming that the measured $(K_La)_{20}$ in the literature is correct, but there will be experimental error associated with that measurement as well. The improvement over the existing model can be as much as 30% since the error of the old model can be as much. It is recommended that this equation replaces the current ASCE 2-06 model.
- Although some extensive properties may change in response to a change in temperature in the new model, such as the volumetric gas flow rate, bubble size, barometric pressure, etc., small changes in these extensive properties can be easily normalized within a reasonable temperature range, such as in the

treatment of Hunter's data, where the gas flow rates are normalized, resulting in an improved correlation. However, this should be verified with testing before changing the standard. It may be difficult to normalize some extensive variables, such as the rotating speed of an impeller-sparger type of aeration system. The effect of such extensive variables has not been discussed in this manuscript, and if normalization is impractical, testing is required in the same way that the theta model would require.

- Hunter's assertion that "equations do not exist... for full scale aeration systems that express K_La as a function of G [the temporal velocity gradient which is dependent on viscosity]" is incorrect. This is because even though turbulence affects K_La substantially, the new 5th power model has excluded the turbulence effect due to temperature. Therefore, as long as such extensive variables are fixed, any one test result can be extrapolated to estimate $(K_La)_{20}$.

- The discovery of a 5th order relationship between solubility and temperature leads to the hypothesis that solubility (C_S) is related to (K_La), but in an inverse manner. Since solubility is related to pressure as given by Henry's Law, K_La must be related to pressure as well. Therefore, in a clean water test with a deep tank, the effect of pressure at the equilibrium level may need to be considered in the use of (Eq. 2.1) for the temperature correction model on K_La.

- Apart from the advantage of a more accurate prediction of $(K_La)_{20}$, the temperature correction model has advantages in design. In a treatment process, the best design is usually when the oxygen consumption balances the oxygen supply. This balance is needed not only to save energy but is also beneficial from the standpoint of the welfare of the microorganisms which are very sensitive to water temperature. It is seldom practical to conduct full scale testing for a range of water temperatures under process conditions. Therefore, a more accurate prediction of K_La_T would enhance designing the treatment process. This is certainly an enormous advantage in the application of equation CG-1 in ASCE 2-06 for designing the oxygen transfer rate in process condition. However, wastewater characteristics do need to be individually measured in the application of the proposed model, since the properties of the fluid E, ρ and σ may all be different from that of pure water or tap water.

REFERENCES

Andrade, Julia. (2013). Solubility Calculations for Hydraulic Gas Compressors. Mirarco Mining Innovation Research Report.

ASCE/EWRI 2-06. (2007). Measurement of Oxygen Transfer in Clean Water. ASCE Standard. ISBN-13: 978-0-7844-0848-3, ISBN-10: 0-7844-0848-3, TD458.M42 2007.

ASCE-18-96. (1997). Standard Guidelines for In-Process Oxygen Transfer Testing. ASCE Standard. ISBN-0-7844-0114-4, TD758.S73 1997.

Baillod, C.R. (1979). Review of oxygen transfer model refinements and data interpretation. Proc., Workshop toward an Oxygen Transfer Standard, U.S. EPA/600-9-78-021, W.C. Boyle, ed., U.S. EPA, Cincinnati, 17-26.

Benson, Bruce B., Krause, Daniel Jr. (1984). The concentration and isotopic fractionation of oxygen dissolved in freshwater and seawater in equilibrium with the atmosphere. Department of Physics, Amherst College, Amherst, Massachusetts 0 1002.

Boogerd, F.C., Bos, P., Kuenen, J.G., Heijnen, J.J., Van der Lans, R.G.J.M. (1990). Oxygen and carbon dioxide mass transfer and the aerobic, autotrophic cultivation of moderate and extreme thermophiles: a case study related to the microbial desulfurization of coal. Biotechnology and Bioengineering 35: 1111-1119. DOI: 10.1002/bit.260351106.

Engineering ToolBox. (2014). Solubility of oxygen in equilibration with air in fresh and sea (salt) water pressures ranging 1-4 bar abs. [online] Available at: https://www.engineeringtoolbox.com.

Engineering Toolbox. (2016). Solubility of oxygen in equilibration with air in fresh and sea (salt) water pressures ranging 1-4 bar abs. [online] Available at: <http://www.engineeringtoolbox.com/> [Accessed 8 May 2016].

Haslam, R.T., Hershey, R.L., Kean, R.H. (1924). Effect of gas velocity and temperature on rate of absorption. Industrial and Engineering Chemistry 16(12): 1224-1230. DOI: 10.1021/ie50180a004 Publication Date: December 1924.

Hunter, John S. (1977). A Basis for Aeration Design. Doctor of Philosophy Dissertation, Department of Civil Engineering, Colorado State University, Fort Collins CO.

Hunter, John S. III. (1979). Accounting for the Effects of Water Temperature in Aerator Test Procedures. EPA Proceedings Workshop Toward an Oxygen Transfer Standard EPA-600/9-78-021.

IAPWS The International Association for the Properties of Water and Steam, Berlin, Germany September 2008, "Release on the IAPWS Formulation 2008 for the Viscosity of Ordinary Water Substance", 2008 International Association for the Properties of Water and Steam Publication. (September 2008).

Lee, J. (1978). Interpretation of Non-steady State Submerged Bubble Oxygen Transfer Data. Independent study report in partial fulfillment of the requirements for the degree of Master of Science (Civil and Environmental Engineering) at the University of Wisconsin-Madison, 1978 [Unpublished].

Lee, J. (2017). Development of a model to determine mass transfer coefficient and oxygen solubility in bioreactors, Heliyon, Volume 3, Issue 2, February 2017, e00248, ISSN 2405-8440. http://doi.org/10.1016/j.heliyon.2017.e00248.

Lee, J. (2018). Development of a model to determine the baseline mass transfer coefficients in aeration tanks. Water Environment Research 90(12): 2126.

Lewis, W.K., Whitman, W.G. (1924). Principles of gas absorption. Industrial and Engineering Chemistry 16(12): 1215-1220. Publication Date: December 1924 (Article) DOI: 10.1021/ie50180a002.

Metcalf & Eddy, Inc. second edition. Wastewater Engineering: Treatment & Disposal. ISBN 0-07-041677-X.

Rounds, Stewart. (2011). Technical Memorandum, Office of Water Quality Technical Memorandum 2011.03. "Change to Solubility Equations for Oxygen in Water", USGS Oregon Water Science Center.

Tromans, D. (1998). Temperature and pressure dependent solubility of oxygen in water: a thermodynamic analysis. University of British Columbia Department of Metals and Materials Engineering, 6350 Stores Road, Vancouver, British Columbia Canada, V6T 1Z4, Hydrometallurgy 48 _1998. 327-342.

Vogelaar, J.C.T., Klapwijk, A., Van Lier J.B., Rulkens, W.H. (2000). Temperature Effects on the Oxygen Transfer Rate between 20 and 55°C. Water Research Vol. 34, No. 3, Elsevier Science Ltd.

3

Development of a Model to Determine Baseline Mass Transfer Coefficients in Aeration Tanks

3.0   INTRODUCTION

The term $(K_L a)$ has been widely used to mean the mass transfer coefficient of both micro-scale and macro-scale aeration. On a macro-scale, the efficiency of porous fine-bubble diffusers varies from 10 to 30 percent or more, depending on tank depth. Due to the uncertainty in predicting the efficiency of an aeration system, the actual oxygen requirement cannot be accurately determined, even though air use is a key parameter in sizingblowers, air piping, and the number of diffuser plants in treatment plant aeration design. A safety factor as high as 2 is sometimes assigned to compute the actual oxygen requirement in the sizing of blowers (Metcalf and Eddy 1985). Furthermore, it has been observed by Eckenfelder (1952) and other researchers that the Standard Oxygen Transfer Rate (SOTR) is constant for a given aeration system, and that at different temperatures $K_L a$ and C_S (saturation concentration or oxygen solubility) adjust accordingly, giving rise to the conjecture that these two parameters are inversely proportional to each other under certain conditions, which has been proven valid for shallow tanks such as pilot plants or experimental vessels in laboratories (Boogerd et al. 1990, Eckenfelder 1952, Hunter 1979, Vogelaar et al. 2000). This may not be true for deeper tanks.

The objective of this chapter is to present an experimentally validated, mechanistic depth-correction model for fine bubble aeration for different water depths under **the same average volumetric gas flow rate supply** (Q_a), and the same horizontal cross-sectional surface area, so that it can be used to predict the mass transfer coefficient and the SOTE (Standard Oxygen Transfer Efficiency) for aeration tanks. Furthermore, this manuscript shows that the proposed temperature correction model (Lee 2017) as given in Chapter 2 can be applied to non-shallow tanks as well when $K_L a$ is corrected to zero depth using the depth correction model. This depth correction model deals with the changes in the mass transfer coefficient with depth inasmuch as the temperature correction model deals with its changes with temperature. The proposed equations [Eq. (3.6) to Eq. (3-10)] allow calculation of the baseline $(K_L a_0)$ by solving them simultaneously for $K_L a_0$. The baseline coefficient $K_L a_0$ is a hypothetical parameter which is defined in this manuscript as the oxygen transfer rate coefficient at zero depth. The work presented here has shown that the standard baseline $(K_L a_0)_{20}$ (the standardized $K_L a_0$) determined from a single clean water test [ASCE 2007], at any temperature, can predict $(K_L a)_{20}$ (the standardized **bulk liquid** apparent mass transfer coefficient for clean water) for any other tank depth (if the gas flowrate Q_a is kept constant or if the baseline $K_L a_0$ can be normalized to Q_a), using the proposed depth correction model and the temperature correction model together. This manuscript shows that the variation of $K_L a$ with depth is an exponential function with respect to the baseline parameter $K_L a_0$, and this relationship allows a more precise determination of aeration system efficiencies.

3.1 BACKGROUND

Because of the complexity of the subject, a dedicated Chapter 4 seeks to clarify about the progression of the development of the Lee-Baillod model and about the new things beyond the Lee-Baillod model. Specifically, the development of the final model (a set of equations) that would define the baseline $K_L a_0$ occurred in three distinct phases:

Phase 1: This happened in the 70s. Lee [1978] and Baillod [1979] jointly developed Eq. (4.1) to Eq. (4.16) when it was recognized that the saturation concentration C_∞^* is a variable rather than a constant assumed to be so in the oxygen transfer equation. The final equation in this phase Eq. (4.16) was also attempted by Lakin and Salzman (1977), and more recently by McGinnis and Little (2002). This equation exists in a differential form. McGinnis and Little (2002) managed to solve this equation by numerical integration, but their results appeared to have an error of around 15% compared to measured data.

Phase 2: This pertains to Eq. (4.17) to Eq. (4.33). The final equation in this phase was derived by Lee [1978] and Baillod [1979], as well as by Lakin and Salzman [1977], but the latter's equation contains an apparent error in sign [Lee 1978]. Even though Eq. (4.33) was mathematically correct, the rising bubbles according to this model give an ever-increasing mole fraction (it should be the other way round) and so this mole-fraction model was deemed to be unrealistic, but it served as a conveyance equation for integrating this equation into a practical oxygen transfer equation, allowing the researchers to conclude that the conventional oxygen transfer equation describing the macroscopic transfer model in an aeration tank is valid, when the $K_L a$ is interpreted as the apparent mass transfer coefficient. Dr. Baillod went on to develop a parameter that he called the 'true' $K_L a$ [EPA-600/2-83-102].

Phase 3: The remaining equations Eq. (4.34) to Eq. (4.76) pertain to this phase. The concept of 'true' $K_L a$ was not accepted as valid by the Standards Committee [ASCE 2007], and rightly so. There are at least two reasons why this concept cannot be right:

First of all, the standard transfer equation given by $dc/dt = K_L a (C_\infty^* - c)$ is correct. In a majority of non-steady state clean water tests, this model never fails to give a very good fit to the re-aeration data;

Secondly, the calculated 'true' $K_L a$ is always higher than the apparent $K_L a$. If the parameter estimation has under-estimated $K_L a$, then it must also have over-estimated C_∞^*, since the two are co-related. Jiang and Stenstrom [2012] have monitored the off-gas content in non-steady state clean water tests, and it was shown that the oxygen in the off-gas is depleted in the early part of the test and then returns to 0.2095 mole fraction at the end of the test. This means there is no net transfer when the system has reached the steady state. Had the saturation concentration been over estimated, one would expect that there would still have been net negative transfer even at steady state, and the exit mole fraction would have exceeded 0.2095 if the average saturation concentration had been less than the bulk average DO concentration at equilibrium. This obviously had not happened and would not have been logical. In other words, the measured C_∞^* must be the true saturation concentration and must not be overestimated. Since $K_L a$ and C_∞^* are related inversely to each other, the measured $K_L a$ must also be the true $K_L a$ and not under-estimated as well.

So, what does this 'true' $K_L a$ parameter mean?

Consider the infinite series for any variable x:

$$e^x = 1 + x + \frac{x^2}{2!} + \frac{x^3}{3!} + \frac{x^4}{4!} + \dots + \frac{x^n}{n!} + \dots$$

In Chapter 4, the proposed model is (see Chapter 4) that the proposed model is given by:

$$K_L a = \frac{1 - \exp(-K_L a_0 \Phi Z_d)}{\Phi Z_d}$$

where

$$\Phi = [HRST/Q_a] (1 - e)$$

Since, by expanding the exponential function into a series,

$$\exp(-K_L a_0 \Phi Z_d) = 1 + (-K_L a_0 \Phi Z_d) + \frac{(-K_L a_0 \Phi Z_d)^2}{2} + \frac{(-K_L a_0 \Phi Z_d)^3}{3!} + \dots$$

Hence,

$$K_L a = \frac{1 - \left[1 + (-K_L a_0 \Phi Z_d) + \frac{(-K_L a_0 \Phi Z_d)^2}{2} + \frac{(-K_L a_0 \Phi Z_d)^3}{3!} + \dots \right]}{\Phi Z_d}$$

Therefore,

$$K_L a = K_L a_0 - K_L a_0^2 \Phi Z_d / 2 + K_L a_0^3 (\Phi Z_d)^2 / 3! \dots$$

Thus, when Z_d tends to zero, $K_L a$ tends to be $K_L a_0$.

Therefore, it was found by the author that the so-called 'true' $K_L a$ is a link between the apparent $K_L a$ and the surface $K_L a$, where the phenomenon of gas-side depletion is eliminated. By adjusting the equations for both parameters (C_∞^* and $K_L a$) with calibration factors, based on the effective depth (de) or the effective depth ratio (e), the proposed model linking $K_L a$ and $K_L a_0$ became valid for certain conditions, using existing data to verify. The model is further enhanced by recognizing that, at saturation, the mole-fraction variation curve is a concave curve, so that there is a minimum point in this curve that corresponds to a minimum oxygen mole fraction, at which the absorption rate and the desorption rate are equalized at equilibrium. The derivative (dy/dz) at this point must therefore equal zero.

The so-called 'true' $K_L a$ is in fact a parameter representing the mass transfer coefficient when the gas depletion is absent (i.e. tank height of zero). This can be verified by plotting this baseline $K_L a_0$ against the inverse of oxygen solubility in water (handbook values) (see Fig. 3.5), and one would expect a straight line passing through the origin, because when the tank height is *infinitesimally* small, $K_L a$ becomes $K_L a_0$ and C_∞^* reduces to C_S. In Chapter 2, it has been explained that these two entities are inversely related to each other. This explains why, for surface aeration in shallow tanks where oxygen is derived from atmospheric air rather than from diffused submerged bubbles, the inverse relationship between $K_L a$ and C_S would hold because gas depletion does not exist in such cases. It also explains why $K_L a_0$ is always higher in value than $K_L a$ as seen in Fig. 3.8, since gas depletion hinders gas transfer. $K_L a_0$ without depletion must therefore be higher than that with depletion. This does not mean that the hypothetical oxygen transfer rate at zero depth would become higher, since at the surface, the saturation concentration is simultaneously smaller than that in the bulk liquid. However, it is generally accepted that the deeper the tank, the higher the oxygen transfer efficiency, all things else being equal [Houck and Boon 1980] [Yunt and Hancuff 1988a, 1988b]. The suite of equations entailing this phase of the modeling also takes care of the hydraulic pressure variations with respect to depth, so that rising bubbles would experience volume changes, even without the gas depletion, due to this hydrostatic phenomenon.

3.2 MODEL DEVELOPMENT

While C_S is proportional to pressure as stated by Henry's Law, and the effect of hydrostatic pressure on the dissolved oxygen (DO) saturation concentration is linear with changing water depths, its effect on $K_L a$ is less certain. Boon (1979) found that the effect of immersion depth on $K_L a$ is a general decline in the $K_L a$ versus increasing depths. This holds for different equipment configuration and for different tank shapes. However, beyond a certain depth, this declining trend ceases and the $K_L a$ value descends to a constant value.

The relationship between $K_L a$ and depth is not linear and is different from the relationship between C_S (here C_S is used in the context of equilibrium saturation concentration in a bulk liquid, better known as C_∞^*) and depth, which is always an increasing function with depth. Houck and Boon (1980) found that the aeration efficiency should improve with increasing tank depth, but their data showed no clear correlation between tank depth and oxygen efficiency at depths greater than 3.6 m (12 ft). Another interesting observation is that the variations of the $K_L a$ with depth is dependent on the gas flow rates. Furthermore, their studies show that increases in blower efficiency can be expected up to about 9.1 m (30 ft). Oxygen depletion beyond this depth clouds their analysis.

Therefore, it is not likely that the inverse proportionality between $K_L a$ and C_S still holds for deep tanks, so that the previous findings of the researchers may be overturned for deep tank aeration. In ASCE 18-96 Standard Guidelines (ASCE 1997), it is stated that the traditional temperature correction coefficient (θ) for $K_L a$ offsets the corresponding temperature correction coefficient for the saturation concentration of DO in water. However, K_La is not only a non-linear function of depth but also a function of a host of other factors (e.g. temperature, gas flow rate, superficial velocity U_g–the unit average gas flow rate over the cross-sectional area of the tank, mixing intensity, etc.) (Metzger 1968). The data from the literature describe the general tendency of the two variables (K_La and C_S) to move in different directions when temperature is changed, but the product is not constant for *non-shallow* tanks. The statement in the Standard Guidelines is questionable for deep tanks or any tank with a significant physical height. Lee (2017) showed that for $P_S = 1$ atm, corresponding to a negligible tank depth, and for a constant average volumetric gas flow rate Q_a (m^3/min), K_La is directly proportional to the water properties (this finding is also supported by Daniil and Gulliver (1988)), as well as to the 5th power of temperature in Kelvin, as shown in Chapter 2 Eq. (2.1) and again shown by Eq. (3.1) below:

$$(K_La)_T = K \times T^5 \times \frac{E\rho\sigma}{P_S} \tag{3.1}$$

where

K is a proportionality constant;

K_La_T (min^{-1}) is mass transfer coefficient at T;

T is any temperature in K (Kelvin);

E is modulus of elasticity of water in (kN/m^2) $\times 10^{-6}$;

ρ is density of water (kg/m^3);

σ is surface tension of water (N/m);

P_S is the saturation pressure in atmospheres (atm).

Based on the experiments by Hunter (1979) and Vogelaar et al. (2000), Eq. [3.1] would supersede the traditional Arrhenius equation for temperature correction because of the higher accuracy, especially for water temperatures above 20°C. Lee (2017) hypothesized that K_La is inversely proportional to P_S, but this applies only to shallow tanks, according to those experiments.

The variation of the mass transfer coefficient K_La at different depths is due to the phenomenon known as gas depletion. Gas-side depletion refers to the decrease in oxygen partial pressure as the bubbles rise through the water column and is the major mechanism for oxygen transfer in submerged bubble aeration. As the air bubbles rise, oxygen is transferred but no net nitrogen is transferred. This occurs because the nitrogen concentration in the tank column is constant, since there are no reactions that consume dissolved nitrogen [Stenstrom 2001]. All diffused aeration systems will experience higher gas-side depletion as the water depth increases because of the longer contact times of the bubbles with water.

According to Stenstrom (2001), more efficient systems encounter gas-side oxygen depletion at shallow depths. Coarse bubble diffusers may not experience gas-side depletion until 15 m (50 feet) or more of depth. Fine pore systems experience gas side depletion at shallower depths, but typically not less than 6 m (20 feet). (This statement appears to be incorrect. The data from Yunt and Hancuff [1988a] seems to suggest that gas-side depletion is significant even at 3 m (10 feet) for diffused aeration. In fact, without gas depletion, there would be no gas transfer except any transfer from the open atmosphere). Fine pore diffuser systems using full floor coverage typically have standard transfer efficiencies from 2 to 2.5% per cent per 0.3 m (1 foot) (SOTE/0.3 m), depending on the gas flow rate and diffuser density. For systems of high SOTE/ 0.3 m, it is not surprising that gas side depletion occurs at shallow depths. (Again, this usage of SOTE per unit depth is not justified as the variations with depth is not linear). However, at the baseline of zero depth, there would be no gas depletion, and Eq. (3.1) for temperature correction should apply.

The model developed to calculate the baseline $(K_L a_0)$ was based on fundamental gas-liquid gas transfer principles and considered the oxygen mass balances on a rising bubble of a constant volume, leading to the validation of the basic model for the non-steady state clean water test as described in the ASCE 2-06 standard [ASCE 2007]. During the validation, two important models have been found-the depth correction model that gives a meaning to the apparent $K_L a$ in relation to the liquid depth, and the Lee-Baillod model (as defined by the author herewith) that describes the relationships between equilibrium concentration, depth and exit gas composition, based on an oxygen mole fraction variation curve. The mathematical derivation of the Lee-Baillod model based on a mass balance in the gas phase is given in Chapter 4. This gives rise to an expression Eq. (4.33) as shown below:

$$y = \frac{C}{HP} + \left(\frac{Y_0 P_d}{P} - \frac{C}{HP} \right) \exp(-\Omega z)$$

where y is the oxygen mole fraction at any point in the aeration tank as the bubble ascends to the water free surface; C is the dissolved oxygen concentration in the bulk liquid; H is Henry's Law constant; Y_0 is the mole fraction at the initial bubble release; P_d is the pressure at the diffuser; P is the absolute pressure at the point corresponding to z; z is the depth measured from the bottom (meter); Ω is a constant (see Eq. 4.25). As for the liquid phase mass balance, the net accumulation rate of dissolved gas in the liquid column is equal to the gas mass flow rate delivered by diffusion, if there is no gas escape from the liquid column. The details of the derivation arising from mass balance of the liquid phase is given in Section 4.1.2.

Therefore, from the mass balances in a non-steady state clean water test,

$$\frac{dC}{dt} = K_1(K_2 - C) \tag{3.2}$$

where

$$K_1 = K_L a_0 \frac{(1 - \exp(-\Omega Z_d))}{\Omega Z_d} \tag{3.3}$$

and

$$K_2 = HY_0 P_d \tag{3.4}$$

Thus, the basic transfer equation (the Standard Model) is proven mathematically, since K_1 has the same meaning as $K_L a$, and K_2 has the meaning of the saturation concentration C_∞^*. Eq. 3.3 is the **depth correction model** for the $K_L a$. However, as defined in Chapter 4, the Lee-Baillod model, Eq. [4.33], that calculates the oxygen mole fraction y, is not physically correct because of the inherent assumptions, particularly the constant bubble volume assumption. However, it can be amended to Eq. [3.5] below by inserting two parameters n and m where appropriate, that is, based on a variable oxygen mole fraction curve versus the tank depth, as can be seen in Fig. 3.1 (showing the case when the DO is approaching saturation concentration C_∞^*). This equation is different from other developed models such as the Downing-Boon's model [Downing and Boon 1968] and model developed by Jackson and Shen (1978), which are linear. Most models predict the equilibrium level or the saturation level to be located at mid-depth ($d_e/Z_d = 0.5$) which is unrealistic as both the mole fraction and the bubble interfacial area change during the bubble rise to the surface, so that $d_e/Z_d = <0.5$, where e is the effective depth ratio given by d_e/Z_d, where d_e is the effective depth, and Z_d is the immersion depth of diffuser. After modification of the CBVM (Constant Bubble Volume Model) with the calibration factors 'n' and 'm' for the Lee-Baillod model (Eq. (4.33)], to account for the *non-constant* bubble volume in a deep tank, the following generalized equation is obtained:

$$y = \frac{C}{nHP} + \left(\frac{Y_0 P_d}{P} - \frac{C}{nHP}\right)\exp(-x \cdot K_L a_0 \cdot mz) \tag{3.5}$$

where m and n are calibration parameters for the curve; $Hk = \Omega$. (It can be shown that $\Omega = x \cdot K_L a_0$, where $x = HRT/U_g$, where U_g is the height-averaged superficial gas velocity; R is specific gas constant for oxygen; H is Henry's constant; T is absolute temperature). Other symbols are as shown in Fig. 3.1 below. This equation is equivalent to Eq. 4.58 in Section 4.1.5 of Chapter 4.

This equation for the generalized Lee-Baillod model can be differentiated by calculus with respect to z, and then setting it to zero to obtain the minimum point. Another equation is thus developed that gives the point along the curve at which the minimum mole fraction of oxygen occurs. Similarly, the modified equation (Eq. 3.5) can be subjected to mathematical integration just like the previous case for the constant bubble volume model. All the resulting equations that lend themselves to five simultaneous equations for solving the unknown parameters (n, m, $K_L a_0$, y_e, Z_e) are summarized below:

$$K_L a = \frac{[1 - \exp(-K_L a_0 x(1-e)Z_d)]}{x(1-e)Z_d} \tag{3.6}$$

$$y = \frac{C}{nHP} + \left(\frac{Y_0 P_d}{P} - \frac{C}{nHP}\right)\exp(-x K_L a_0 \cdot mz) \tag{3.7}$$

$$C_\infty^* = nH \times 0.2095 \times \frac{P_a - P_d \exp(-mx \cdot K_L a_0 \cdot Z_d)}{1 - \exp(-mx \cdot K_L a_0 \cdot Z_d)} \tag{3.8}$$

$$K_L a = \frac{1 - \exp(-mx \cdot K_L a_0 \cdot Z_d)}{nmx \cdot Z_d} + \frac{(n-1)K_L a_0}{n} \tag{3.9}$$

$$Z_e = \frac{1}{mx K_L a_0}\left\{\ln\left(P_e \frac{mx K_L a_0}{nrw}\right) + \ln\left(\frac{nHY_0 P_d}{C_\infty^*} - 1\right)\right\} \tag{3.10}$$

The above equations are identical to Eqs. 4.48, 4.58, 4.63, 4.65, 4.74 in the next chapter.

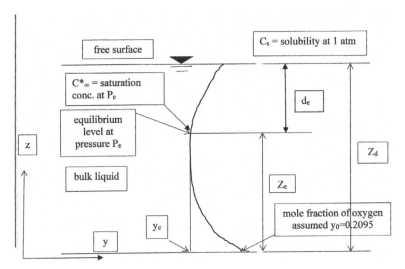

FIGURE 3.1 The MF (mole fraction) curve for the Lee-Baillod model (subscript e=equilibrium)

Derivation of the above equations is given in Chapter 4. These equations are repeated in the calculation sheet (Table 3.2) below. In the above suite of equations, Eq. (3.6) was uniquely derived from the depth correction model, via an adjustment to the effective saturation depth ratio (e) from $e = 1$ to $e = d_e/Z_d$, in the same way the Lee-Baillod model was adjusted by the calibration factors (n, m) to account for the constant bubble volume assumption.

After calculating the baseline mass transfer coefficient $K_L a_0$ at any test temperature T, the standard baseline can be calculated by the use of the proposed 5th power temperature correction model (Lee 2017). The use of the 5th power model as given by Eq. (3.1) is preferred to the current ASCE model (ASCE 2007). The topic of temperature correction is discussed in Section 5.6 in Chapter 5, where various temperature models are compared as shown in the bar graph of Fig. 5.11. The discrepancies between these models in terms of standardizing $(K_L a_0)_T$ to $(K_L a_0)_{20}$ in this exercise are small because all the tests cited were done within the narrow temperature range of 10°C and 30°C [ASCE 2007].

However, the use of a 5th power model appears to give the best regression analysis result to yield the standard baseline $(K_L a_0)_{20}$, but since the temperature effect compared to the depth effect is small, other models will also give similar result, even though the prediction would not be as accurate, especially for water temperatures above 20°C (Lee 2017).

3.3 MATERIAL AND METHOD

$K_L a_0$ can be calculated from a set of clean water test results. The following development is based on data extracted from Yunt and Hancuff (1988a). The test facility used for all tests was an all steel rectangular aeration tank located at the Los Angeles County Sanitation Districts (LACSD) Joint Water Pollution Control Plant. The dimensions of this tank are 6.1 m × 6.1 m × 7.6 m (20 ft × 20 ft × 25 ft) side water depth (SWD). As reported, more than 100 tests had been carried out on various submerged aeration systems in the Control Plant. The test temperatures were reported to be within the range of 16.2°C to 25.2°C. In the aeration tests, multiple diffusers were placed at a submerged depth of 3.05-7.62 m with a tank water surface of 37.2 m^2 and water volume of 113.2-283.1 m^3. Fine bubble diffusers were operated at air flow rates of 213.0-683.4 scmh (standard cubic meters per hour). The FMC diffusers are a fine bubble tube diffuser system manufactured by FMC Corporation. The testing configuration is given in *Figure 10 of the LACSD report*. The diffuser media was a white porous modified acrylonitrile-styrene copolymer material. These tube diffusers had a permeability of 23.7 L/s or 50 scfm (standard cubic feet per minute) at a headloss of 25.4 mm (1 in.) of water. The diffuser air release point was 65 cm (25 in.) above the tank floor. The tests were carried out at four different depths for a range of air flow rates for each depth. The first group of tests were carried out from August 29, 1978 to September 29, 1978 with the last date for one test only. This group is assumed to have a temperature of 25°C for data interpretation and analysis. This group entails the 3.05 m (10 ft) [$Z_d = 2.44$ m] and 7.62 m (25 ft) [$Z_d = 7.02$ m] tanks. The second group was carried out from February 8, 1979 to February 9, 1979 and this group should have a temperature of 16°C. These tanks are 4.57 m (15 ft) [$Z_d = 4$ m] and 6.10 m (20 ft) [$Z_d = 5.6$ m]. Only standard values $(K_L a)_{20}$ and $C^*_{\infty\,20}$ were reported; the corresponding $K_L a_T$ and $C^*_{\infty T}$ were back-calculated from the formulae reportedly used in the conversion to standard conditions:

$$(K_L a)_T = (K_L a)_{20}\theta^{T-20} \tag{3.11}$$

$$C^*_0 = 8.96 + .271Z_{emd} \tag{3.12}$$

where the reported value of θ used was 1.024 and Z_{emd} is the effective depth similar to *(de)* in this development. C^*_0 is identical to $C^*_{\infty\,20}$. The actual measured saturation concentration $C^*_{\infty T}$ can then be back-calculated by another equation using Z_{emd} as an independent variable [Yunt and Hancuff 1988a]. With these parameters thus determined, the baseline $K_L a (K_L a_0)$ for every test can then be determined. For each water depth tested, the volumetric mass transfer coefficients can then be plotted against the average flow rates for both the apparent and the baseline $K_L a$ values. The test results are given in the *LACSD report Table 5*: "Summary of Exponential Method Results: FMC Fine Bubble Tube Diffusers" and copied herewith as Table 3.1 below.

3.4   RESULTS AND DISCUSSIONS

Tables 3.1 and 3.2 below are compiled based on data contained in the LACSD report for FMC Fine Bubble Tube Diffusers. Table 3.1 shows all the raw data as given in the LACSD report. Table 3.2 (Excel spreadsheet for estimating variables $K_L a_0$, n, m, d_e and y_e) shows an example calculation of the baseline mass transfer coefficient $(K_L a_0)_T$ using the Excel Solver with the model equations [Eq. (3.6) to Eq. (3.10)] incorporated for tank 1 run 1. Table 3.3 showcases calculation of simulation tank for the 7.6 m (25 ft) tank at a gas flow rate of 7.96 Nm^3/min (281 scfm) using the same specific baseline as calculated from Table 3.2, and using the same set of developed equations, but with $(K_L a)_{20}$ as the unknown parameter to be solved for.

TABLE 3.1 LACSD (Los Angeles County Sanitation District) report test data (1978) for the FMC diffuser

Date	Rn No.	Water Depth Z	Delivered Power Density	Air-flow Rate Q_s	Temperature	Apparent $(K_La)_{20}$	Apparent $(K_La)_{20}$	Saturation conc. $C_{\infty 20}$	Standard Oxygen Transfer Efficiency
		m	(hp/1000 ft³)	(scmh)	$T(°C)^*$	(1/hr)	(1/min)	(mg/L)	(%)
Aug 29,78	1	3.05	2.02	700	25.2	**17.46**	0.2910	**9.87**	10.06
Aug 29,78	2	3.05	1.16	470	25.2	**13.37**	0.2228	**9.99**	11.68
Aug 29,78	3	3.05	0.54	241	25.2	**7.63**	0.1272	**10.05**	12.95
Aug 29,78	1	7.62	1.66	704	25.2	**14.99**	0.2498	**11.23**	23.93
Aug 29,78	2	7.62	1.07	478	25.2	**11.12**	0.1853	**11.26**	24.40
Aug 29,78	3	7.62	0.51	236	25.2	**6.39**	0.1065	**11.54**	31.71
Sep 29,78	1	3.05	1.19	472	25.2	**13.39**	0.2232	**9.98**	11.61
Feb 08,79	1	4.57	1.81	694	16.2	**16.61**	0.2768	**10.50**	15.34
Feb 08,79	2	4.57	1.05	449	16.2	**11.90**	0.1983	**10.54**	17.07
Feb 08,79	3	4.57	0.51	231	16.2	**6.88**	0.1147	**10.63**	19.87
Feb 08,79	1	6.10	1.74	709	16.2	**16.73**	0.2788	**10.80**	20.69
Feb 08,79	2	6.10	1.08	471	16.2	**11.62**	0.1937	**11.05**	22.17
Feb 08,79	3	6.10	0.49	224	16.2	**6.10**	0.1017	**11.19**	25.04

* **Note:** Water temperature was deduced from the report statement: "The temperature range used in the study was 16.2 to 25.2°C." Reported main data are given in bold; $(K_La)_{20}$ given in this table is based on the Arrhenius model using $\theta = 1.024$ [ASCE 2007]
The 5th power temperature correction model [Lee 2017] to convert K_La_0 estimated in Table 3.2 to $(K_La_0)_{20}$ and subsequently to $(K_La)_{20}$ is given by:

$$(K_La)_{20} = K_La \frac{(E\rho\sigma)_{20}}{(E\rho\sigma)_T}\left(\frac{T_{20}}{T}\right)^5$$

Note that there were some discrepancies in the reported data, in that the data for the 7 m tank, in that the data for the SOTE% were calculated by an equation in the report and they did not match up for two points in the report. These data were discarded and the calculated values using the report's equations were used in the above table, but these data are still suspect. The greater the number of tests are done, the better would be the estimation of the unknown parameters.

TABLE 3.2 Calculation of $K_L a_0$ for run 1 for the 3.05 m (10 ft) tank at $T = 25°C$

Fixed Parameters					For P_e calc.: $C_∞^* = Hy_e P_e$		SS Error
Diffuser depth	$Z_d =$	2.44	m	Saturation depth	de (m) =	1.19	
atm pressure	$P_a =$	101325	N/m²	eff. depth ratio	$e =$	0.49	
Press at diff.	$P_d =$	121964	N/m²	equil. pressure	P_e (N/m²) =	109758	
$x =$	HRST/Q_a	0.1051	min/m	Eq. I = (3.9)	(Eq. I – $K_L a$) =	9.951E-05	9.90288E-09
Tank area	$S =$	37.2	m²	Eq. II = (3.8)	(Eq. II – $C_∞^*$) =	-1.275E-06	1.62751E-12
Variables				Eq. III = (3.10)	(Eq. III – Z_e) =	7.985E-05	6.37732E-09
min^{-1}	$K_L a_0 =$	**0.3349**		Eq. IV = (3.6)	(Eq. IV – $K_L a$) =	5.377E-08	2.89175E-15
Dimensionless	$n =$	5.6629				sum =	1.62818E-08
Dimensionless	$m =$	**3.0737**		Eq. I, II, III, IV, V given below			
Dimensionless	$y_e =$	**0.2080**		off gas = y at exit			
Dimensionless	$y_d =$	0.2095		Eq. V = (3.7)		0.2095	checked
Data							
min^{-1}	$K_L a =$	0.3276					
mg/L	$C_∞^* =$	9.05					
mg/L/(N/m²)	$H =$	3.96E-04					
N/m³	$r_w =$	9777					
Vapor pressure N/m²	$P_{vt} =$	3200					

$$K_L a = \frac{1 - \exp(-mx \cdot K_L a_0 \cdot Z_d)}{nmx \cdot Z_d} + \frac{(n-1)K_L a_0}{n} \tag{3.9}$$

$$C_∞^* = nH \times 0.2095 \times \frac{P_a - P_d \exp(-mx \cdot K_L a_0 \cdot Z_d)}{1 - \exp(-mx \cdot K_L a_0 \cdot Z_d)} \tag{3.8}$$

$$Z_e = \frac{1}{mx K_L a_0}\left[\ln\left(P_e \frac{mx K_L a_0}{nrw}\right) + \ln\left(\frac{nHY_0 P_d}{C_∞^*} - 1\right)\right] \tag{3.10}$$

$$K_L a = \frac{[1 - \exp(-K_L a_0 x(1 - e)Z_d)]}{x(1 - e)Z_d} \tag{3.6}$$

Checking equation at system equilibrium: ($y = y_0 = 0.2095$; $C = C_∞^*$; $z = Z_d$; $P = P_a$):

$$y = \frac{C}{nHP} + \left(\frac{Y_0 P_d}{P} - \frac{C}{nHP}\right)\exp(-x K_L a_0 \cdot mz) \tag{3.7}$$

The simulated result from Table 3.3 gives a value of $(K_L a)_{20} = 0.1874$ min^{-1} as compared to the reported test value of $(K_L a)_{20} = 0.1853$ min^{-1} which gives an error difference of around 1% only.

3.4.1 Example Calculation

For example, suppose the specific baseline $(K_L a_0)_{20}$ has been established by a clean water test to be 4.435×10^{-2} min^{-1} per $Q_a^{0.82}$, where Q_a is in m^3/min. In customary units, it would be 2.38×10^{-3} min^{-1} per $Q_a^{0.82}$ (where Q_a is in cfm). We want to estimate $K_L a$ for a 7.62 m (25 ft) tank with a diffuser submergence 0.6 m (2 ft) above the floor ($Z_d = 7.01$ m), at a gas supply rate of 7.96 Nm^3/min (281 scfm). The horizontal cross-sectional area of the tank is 37.2 m^2 (20 ft × 20 ft). The pressure at the diffuser is given by:

TABLE 3.3 Calculation of $(K_L a)_{20}$ for run 2 for the 7.62 m (25 ft) tank at sp. $(K_L a_0)_{20} = 0.04434$

Simulation of 7.6 m (25-ft) tank at 7.96 m³/min (281 scfm)								
Fixed								SS err
7.6 m	$Z_d =$	7.01			de (m) =	2.95		
20 C	$P_a =$	98992						
	$P_d =$	167613	V(m³) =	283.464	$P_e =$	127855		
	$x =$	0.1957	Q_a (m³/m) =	6.35	Eq. I =	1.16E-03	1.344E-06	
	$S =$	37.2			Eq. II =	−8.37E-06	7.005E-11	
Variables	$K_L a =$	**0.1874**	$(K_L a)_{20} =$	11.25	Eq. III =	2.74E-04	7.499E-08	
	$e =$	**0.42**	sp. $K_L a_0 =$	0.04434	Eq. IV =	−9.04E-04	8.178E-07	
	$n =$	**4.17**			Eq. V =	2.28E-08	5.220E-16	
	$m =$	**2.46**			Eq. VI =	−4.10E-05	1.679E-09	
	$y_e =$	**0.1958**			Min (SS err)		2.236E-06	
	$C_{inf}^* =$	10.91						
Data			$K_L a = \dfrac{1 - \exp(-mx \cdot K_L a_0 \cdot Z_d)}{nmx \cdot Z_d} + \dfrac{(n-1)K_L a_0}{n}$					(Eq. I)
	$K_L a_0 =$	0.2019	$C_\infty^* = nH \times 0.2095 \times \dfrac{P_a - P_d \exp(-mx \cdot K_L a_0 \cdot Z_d)}{1 - \exp(-mx \cdot K_L a_0 \cdot Z_d)}$					(Eq. II)
	$y_d =$	0.2095						
	$H =$	4.383E-04	$Z_e = \dfrac{1}{mx K_L a_0}\left\{ \ln\left(P_e \dfrac{mx K_L a_0}{nrw} \right) + \ln\left(\dfrac{nHY_0 P_d}{C_\infty^*} - 1 \right) \right\}$					(Eq. III)
	$r_w =$	9789						
			$K_L a = \dfrac{[1 - \exp(-K_L a_0 x(1-e)Z_d)]}{x(1-e)Z_d}$					(Eq. IV)
			$y = \dfrac{C}{nHP} + \left(\dfrac{Y_0 nP_d}{nP} - \dfrac{C}{nHP} \right) \exp(-Hk \cdot mz)$					(Eq. V)
	$P_{vt} =$	2333	At system equilibrium ($y = y_0 = 0.2095$; $C = C_\infty^*$; $z = Z_d$; $P = P_a$)					
			$C_\infty^* = \dfrac{(r_w de + P_b - P_{vt})C_{st}^*}{P_s - P_{vt}}$					(Eq. VI)

$$P_d = 101325 + 9789 \times 7.01 = 169946 \text{ N/m}^2.$$

Assuming vapor pressure has no effect on the volumetric gas flowrate, the average gas flow rate is given by (Eq. 2.25) in Chapter 2,

$$Q_a = 172.82\,(293.15)\,(7.96)\,(1/101325 + 1/169946) = 6.35 \text{ m}^3/\text{min (224 cfm)}$$

The superficial velocity $U_g = 6.35/(37.2) = 0.1707$ m/min (0.56 ft/min). Therefore,

$$x = \text{HRT}/U_g = 4.382 \times 10^{-4}\,(0.260)\,(293.15)/0.1707 = 0.1957 \text{ min/m},$$

where R (specific gas constant for oxygen) is given as 0.260 KJ/kg-K; H is the handbook value for Henry's constant at 20°C, given as 4.382E-4 (mg/L)/(N/m²). Assuming $e = 0.45$, $\Phi = 0.1957\,(1 - 0.45) = 0.1076$ min/m (This assumption for 'e' is not needed in the spreadsheet calculations), from Eq. (3.6),

$$K_L a = \frac{1 - \exp(-\Phi Z_d \cdot K_L a_0)}{\Phi Z_d}$$

where $\phi = x\,(1 - e)$

Therefore,

$$K_L a = (1 - \exp\,(-0.1076 \times 7.01 \times 4.435 \times 10^{-2} \times 6.35^{0.82}))/0.1076/7.01$$

$$= 0.1873 \text{ min}^{-1} = 11.23 \text{ hr}^{-1}.$$

This compares with 11.12 hr^{-1} in the real test for this tank. This incurs an error of about +2%

Furthermore, from ASCE 2-06 Eq. (F-1), the effective depth d_e is given as:

$$d_e = \frac{1}{rw}\left[\left(\frac{C_\infty^*}{C_{st}^*}\right)(P_S - P_{vt}) - P_b + P_{vt}\right] \tag{3.13}$$

Rearranging gives,

$$C_\infty^* = \frac{(r_w d_e + P_b - P_{vt})C_{st}^*}{P_S - P_{vt}} \tag{3.14}$$

Therefore,

$$C_\infty^* = (101325 + 9789 \times 0.45 \times 7.01 - 2340) \times 9.09/(101325 - 2340) = 11.92 \text{ mg/L}$$

The above equation (Eq. 3.14) implicitly assumed that the mole fraction at the saturation point is 0.21, but as the Excel Solver calculated, the true mole fraction at equilibrium (Y_e) is 0.1958 (Table 3.3). Therefore, the corrected C_∞^* will be given by:

$$C_\infty^* = (0.1958/0.21) \times 11.92 = 11.11 \text{ mg/L}$$

This compares with the reported measured C_∞^* of 11.26 mg/L. The percent error is about –2%. The calculated SOTR (Standard Oxygen Transfer Rate) is given by (11.23) (11.11) $V = 124.8V$, where V is the volume of tank, which compares well with the SOTR based on reported values of (11.12) (11.26) $V = 125.2V$. The percent error is practically insignificant.

3.4.2 Estimation of the Effective Depth Ratio ($e = d_e/Z_d$)

This paragraph should be read in conjunction with Chapter 4. Figure 3.2 below is a plot of the effective depth ratios calculated from the test runs, the lower line showing the results based on a constant equilibrium mole fraction at 0.21 similar to the equation in ASCE 2-06 Annex F [ASCE 2007], while the top line was based on the Depth Correction Model (Eq. 3.6) and the other developed model equations (see Table 3.2), using Eqs. (4.46) for the effective depth ratio (e); Eq. (4.74) or Eq. (3.10) for Z_e and the minimum Y at Y_e; Eq. (4.72) which is similar to Eq. 3.13 above given by ASCE 2-06 Annex F Eq. (F-1) but corrected for Y_e for calculating d_e; Eq. (4.75) for P_e; Eq. (4.76) for P_a (the atmospheric pressure N/m²). [Eq. 3.8, Eq. 3.9] or [Eqs. (4.63) (4.65)], as derived from the Lee-Baillod model (see Chapter 4) that describes the mole fraction variation curve, and, after inserting boundary conditions, are for calculating C_∞^* and K_La, respectively, which lead to the calibration parameters, n and m; Eq. 3.7 or Eq. (4.58) is used to double check the calculations, as it should give a mole fraction at saturation of 0.2095 at exit.

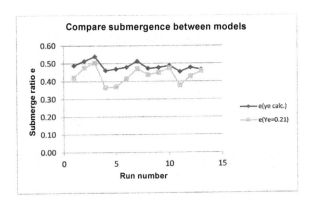

FIGURE 3.2 Comparison of submergence depth ratio (e) rigorous analysis vs. ASCE method

It is important to note that the ASCE 2-06 Annex F Eq. (F-1) has treated Y_e to be the same Y_0 in Eq. (4.72) which is not correct. As a result of this rigorous analysis, the top line in Fig. 3.2 gives a more consistently uniform depth ratio of e.

3.4.3 Determination of the Standard Specific Baseline

Figure 3.3 shows that the resulting K_La_0 values are then adjusted for the standard temperature by the temperature correction equation of the 5th power model (Lee 2017) and plotted against Q_{a20}. Amazingly, all curves fitted together after normalizing K_La_0 values to 20°C, as shown. The exponent is 0.82.

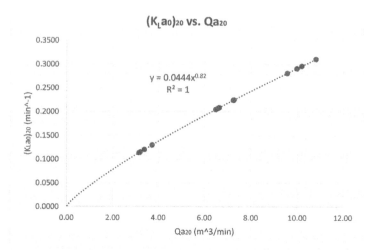

FIGURE 3.3 Standard baseline $(K_La_0)_{20}$ vs. average standard gas flow rate Q_{a20} for various test temperatures

The value obtained from the slope is **44.35 × 10⁻³ (1/min)** for all the gas rates normalized to give the best NLLS (Non-Linear Least Squares) fit, bearing in mind that the K_La_0 is assumed to be related to the gas flowrate by a power curve with an exponent value [Rosso and Stenstrom 2006] [Zhou et al. 2012]. The slope of the curve is defined as the standard specific baseline.

Therefore, the standard specific baseline (sp. $K_La_0)_{20}$ is calculated by the ratio of $(K_La_0)_{20}$ to $Q_a^{0.82}$ or by the slope of the curve in Fig. 3.3. When the same information is compared with a similar plot using the actual measured K_La values (plot not shown), it can be seen that the correlation was still quite good for the curve, but not as exactly as when the baseline values were plotted, testifying the fact that the baseline mass transfer coefficient does represent a standardized performance of the aeration system when the tank is reduced to zero depth (i.e. when the effect of gas depletion in the fine bubble stream was eliminated). As K_La is a local variable dependent on the bubble's tank location, especially its height position, K_La_0 represents the K_La at the surface, i.e. at the top of the tank, where the saturation concentration corresponds to the atmospheric pressure ($P_S = 1$ atm).

3.4.4 Relationship Between the Mass Transfer Coefficient and Saturation Concentration

Figure 3.4 below shows the apparent mass transfer coefficient K_La_T, as reportedly measured, upon normalizing the $(K_La)_T$ values to the air flow rates, plotted against the inverse of the measured saturation concentrations $(C_{\infty T}^*)$ for all the tests, and they give a linear correlation with $R^2 = 0.9859$. This figure shows the inverse relationship between K_La and C_∞^* for different temperatures, but since the K_La data pertain to different gas flow rates (Q_a), K_La must first be normalized to the same Q_a before it can be plotted against C_∞^*, otherwise it would be meaningless because K_La is much more dependent on the air flow rate than the dissolved oxygen saturation concentration would be [Hwang and Stenstrom 1985]. This normalization cannot occur until the relationship between K_La and Q_a is first determined (as shown in Fig. 3.3 above for the baseline plot). Since the parameter estimation based on an assumed power function [Hwang and Stenstrom 1985] [Zhou et al. 2012] has determined that it is a power function of $Q_a^{0.82}$, therefore, the relationship between the specific mass transfer coefficient and the saturation concentration is given by

$$K_La/Q_a^{0.82} = 0.4515\,(1/C_\infty^*),$$

as shown in the graph of Fig. 3.4.

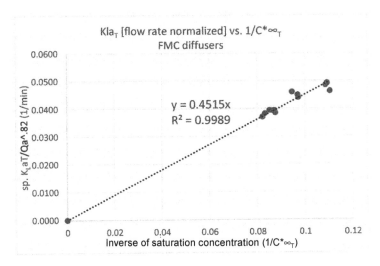

Kla_T [flow rate normalized] vs. $1/C^*\infty_T$
FMC diffusers

$y = 0.4515x$
$R^2 = 0.9989$

sp. $K_LaT/Qa^{.82}$ (1/min)

Inverse of saturation concentration $(1/C^*\infty_T)$

FIGURE 3.4 Specific K_La_T vs. the inverse of measured saturation concentrations $(C_{\infty T}^*)$

As will be seen later, using the same power exponent for tanks other than the baseline is an approximation only, as $R^2 = 0.9859$ is not as good as when the baseline values are used.

3.4.5   Relationship Between the Baseline and Oxygen Solubility

Figure 3.5 shows the relationship between the baseline coefficients ("air flow normalized" base line coefficients, defined as the specific baseline), with the solubility of oxygen in water (C_S), using handbook values [ASCE 2007] for the oxygen solubility. As expected, they bear an inverse correlation, such that when K_La_0 is plotted against the insolubility ($1/C_S$), a linear graph is obtained. Since K_La_0 represents the mass transfer coefficient at the surface and based on the hypothesis that K_La and C_S are inversely proportional to each other [Lee 2017], the specific K_La_0 for all the tanks tested at temperature 16°C would converge into a single point, and all the tanks tested at 25°C would focus onanother single point, regardless of the individual tank depth and the gas flowrate being applied.

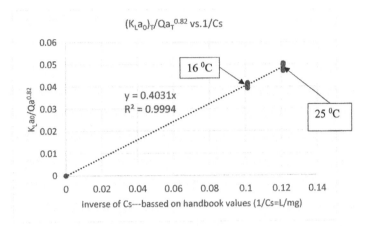

$(K_La_0)_T/Qa_T^{0.82}$ vs.$1/Cs$

FIGURE 3.5 sp. K_La_0 vs. the inverse of surface saturation (solubility) concentration C_S

The specific baseline mass transfer coefficient at standard temperature (20°C) would become a single value, regardless of the tank depth (as shown later by the top curve in Fig. 3.8).

One would therefore expect that K_La_0 would be proportional inversely to the surface saturation value or the solubility when both parameters are changed with temperature [Lee 2017]. The relationship between the specific *baseline* mass transfer coefficient and the gas solubility is given by $K_La_0/Q_a^{0.82} = 0.4031(1/C_S)$, as shown in the graph of Fig. 3.5.

Although the slope of this curve given as $K = 0.4031$ indicates a linear proportionality, the proportionality constant between the baseline and its solubility is not the same as the proportionality constant between K_La_0 and its corresponding temperature function, which is given in Fig. 3.7 where the proportionality constant has a different value, $K = 0.1284$.

3.4.6 Relationship Between the Baseline and the Gas Flow Rate

The equation for converting the gas mass flow rate (Q_S) in col. 5 of Table 3.1 to the average volumetric gas flowrate (Q_a) is given by Eq. (2.25) in Chapter 2. Figure 3.6 below shows the resulting plot of $(K_La_0)_T$ vs. Q_{aT} for the four different tank depths in this experiment.

Two distinct bands of curves were discovered: the 3.05 m (10 ft) [$Z_d = 2.44$ m] tank and the 7.62 m (25 ft) [$Z_d = 7.02$ m] tank were carried out at 25°C forming one band, and the 4.57 m (15 ft) [$Z_d = 4$ m] and the 6.09 m (20 ft) [$Z_d = 5.6$ m] tanks form a different band at 16°C. Here we have the shallowest and the deepest tank curves almost coinciding, and similarly, for the other two tanks, the curves are banded together as another group at 16°C. All the $(K_La_0)_T$ values fitted together at a single temperature indicates that the baseline $(K_La_0)_T$ is quite independent of depth, for the different flow rates at each temperature. They tend to be best fitted by power curves, with the average exponent of around 0.8. This phenomenon is equally pronounced when the apparent K_La values are plotted (not shown), instead of the baselines K_La_0

but with a lesser correlation for each band, especially between the 3.05 m (10 ft) and 7.62 m (25 ft) tank.

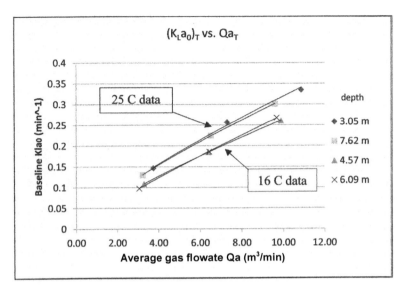

FIGURE 3.6 Baseline K_La_0 vs. average gas flowrate Q_a at various temperatures

3.4.7   Relationship Between the Baseline and Water Temperature

Figure 3.7 below shows the relationship between the baseline mass transfer coefficient $(K_La_0)_T$ and the temperature function based on the 5th power model as given by Eq. (3.1) that was developed in a previous manuscript (Lee 2017) and in Chapter 2.

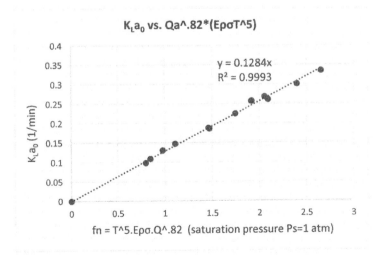

FIGURE 3.7 Baseline $(K_La_0)_T$ plotted against temperature function of 5th power model

Since Eq. [3.1] was based on a constant Q_a, and the relation between K_La_0 and Q_a has been established as power function of 0.82 as shown in Fig. 3.3, and $P_s = 1$ atm for the baseline, the correlation (K_La_0 vs. $Q_a^{0.82}(E\rho\sigma T^5)$) shown in Fig. 3.7 has a value of $R^2 = 0.9958$, testifying that the 5th power model for temperature conversion is valid. The slope of this curve given by $K = 0.1284$ would be identical to the proportionality constant K specified in Eq. [3.1] for the temperature correction model at $P_s = 1$ atm, since K_La_0 pertains to this pressure. Therefore, for this particular case where Q_a is not constant, Eq. 3.1 would become:

$$(K_La_0)_T = 0.1284 \times T^5 \times \frac{(E\rho\sigma)_T}{P_s} \times Q_T^{0.82} \tag{3.15}$$

For comparison, the actual measured mass transfer coefficients K_La are plotted against the same temperature function, and it can be seen that a good correlation can still be obtained, but the correlation is less precise than when the baselines are plotted. Therefore, Eq. 3.1 is still correct for the relationship (as shown in Fig. 3.8) and the temperature correction model, as discussed in Chapter 2. It is believed that the deeper the tank, the further apart from linearity this plot will be, but it appears that the temperature correction model holds for depths up to 7.6 m in this case.

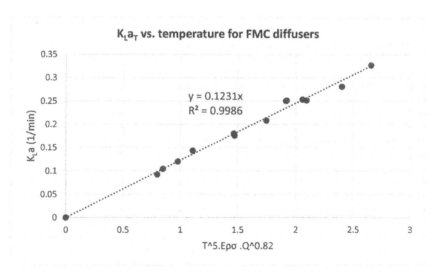

FIGURE 3.8 $(K_La)_T$ plotted against temperature function of the 5th power model

3.5 DISCUSSION AND IMPLICATIONS

Modelling for the mass transfer coefficients through the use of image analysis (as is often the case) of bubble size for diffused aeration is difficult and often system specific with ~15% inaccuracies (McGinnis and Little 2002, Fayolle et al. 2007), especially if measuring in mixed liquor. To avoid using bubble size as an input parameter, a mathematical, mechanistic, model has been developed to predict mass transfer coefficients in deep water tanks when aeration is being performed by the submerged diffused bubble-oxygen transfer mechanism. Using clean water tests data created by Yunt and Hancuff (1988a), the model formula precisely calculates a uniform value of $K_L a_0$ that is independent of tank depth for the standardized $(K_L a_0)_{20}$ at 20°C. This baseline value is equivalent to the surface $K_L a$ that one would obtain from surface aeration, where the effect of gas depletion is negligibly small.

TABLE 3.4 Comparison of predicted and reported clean water tests results

Run no.	Z_d (m)*	T (°C)	rpt. $C^*_{\infty 20}$	rpt. $(K_L a)_{20}$	p. $C^*_{\infty 20}$ (Hye Pe)	p. $(K_L a)_{20}$	rpt. SOTE% (eqt)	p. SOTE%	%err (SOTE)	%err $(K_L a)_{20}$
1	2.44 (8)	25	9.87	0.2910	9.68	0.2993	10.09	10.18	0.9	2.8
2	2.44 (8)	25	9.99	0.2228	9.68	0.2288	11.66	11.60	-0.5	2.6
3	2.44 (8)	25	10.05	0.1272	9.68	0.1305	13.03	12.87	-1.2	2.5
1	7.01 (23)	25	11.23	0.2498	11.56	0.2582	24.51	26.08	6.4**	3.2
2	7.01 (23)	25	11.26	0.1853	11.56	0.1915	26.88	28.52	6.1**	3.2
3	7.01 (23)	25	11.54	0.1065	11.56	0.1097	32.02	33.04	3.2	2.9
1	2.44 (8)	25	9.98	0.2232	9.68	0.2288	11.62	11.56	-0.5	2.5
1	3.96 (13)	16	10.50	0.2768	10.31	0.2759	15.46	15.12	-2.2	-0.4
2	3.96 (13)	16	10.54	0.1983	10.31	0.1976	17.18	16.73	-2.6	-0.4
3	3.96 (13)	16	10.63	0.1147	10.31	0.1141	19.47	18.79	-3.5	-0.5
1	5.49 (18)	16	10.80	0.2788	10.93	0.2787	20.90	21.15	1.2	-0.1
2	5.49 (18)	16	11.05	0.1937	10.93	0.1930	22.34	22.03	-1.4	-0.3
3	5.49 (18)	16	11.19	0.1017	10.93	0.1012	25.00	24.31	-2.8	-0.5

Notes: * numbers in brackets are in feet; diffuser depth Z_d is two feet off the tank floor;
   ** data error, see Table 3.1 footnote; assumptions: $e = 0.48$ and $ye = 0.2$; $P_v = 2333$ N/m^2
Symbols: p. = predicted; rpt. = reported; eqt = equation in Yunt's report.

This chapter has illustrated that, for a set of tanks of different heights subjected to a series of gas flowrates under different temperatures, they yield different values of $(K_L a)_{20}$. However, when all the $(K_L a)_{20}$ values are plotted against their respective flowrates Q_{a20}, a good correlation should be obtained regardless of what the tank heights are. On the other hand, though, when the baseline $(K_L a_0)_{20}$ is plotted against the same Q_{a20} values, an almost perfect power curve correlation is obtained. The simulated results for the $(K_L a)_{20}$ for the various runs in the test are given in Table 3.4 below using the standard specific baseline $(K_L a_0)_{20}$ of **44.35 × 10^{-3} (1/min)** to predict the standard mass transfer coefficients, $(K_L a)_{20}$. The results are then compared with the reported values of the same, given in column 5 (reported) and column 7 (predicted) of the table. The new model using the concept of a baseline $K_L a$ ($K_L a_0$) predicts oxygen transfer coefficients to be within 1~3% error compared to observed measurements and around the same for the standard oxygen transfer efficiency (SOTE%), as shown in Table 3.4 (where p. stands for predicted values and rpt. means the reported values).

It must be remembered that the actual $(K_L a)_{20}$ and $C^*_{\infty 20}$ were never measured at 20°C. The conversion from the test temperature to the standard temperature of 20°C for $K_L a$ in the LACSD report was based on the Arrhenius model that assumed $\theta = 1.024$ which may be the reason the error is larger for the data pertaining to 25°C, since it is known that the temperature model becomes more inaccurate for temperatures higher than 20°C (Lee 2017). Similarly, for the conversion of C^*_∞, the report relied on the ASCE (2007) method, but the barometric pressurewas not reported. The reported values of these standardized parameters may therefore be imprecise and carry some inherent errors in the estimation of the baseline.

3.5.1   Rating Curves for Aeration Equipment

The good prediction of $K_L a$ and the subsequent SOTE (standard oxygen transfer efficiency) is a breakthrough since the correct prediction of the volumetric mass transfer coefficient $(K_L a)$ is a crucial step in the design, operation and scale up of bioreactors including wastewater treatment plant aeration tanks, and the equations developed allow doing so without resorting to multiple full-scale testing for each individual tank under the same testing conditions for different tank heights and temperatures. A family of rating curves for $(K_L a)_{20}$ with respect to depth can thus be constructed for various gas flow rates applied, such as the one shown below (Fig. 3.8). In the chart, the rating curves were constructed based on the three average gas flow rates, the individual flow rates of which vary slightly for each tank depth (see Table 3.1). This has resulted in one tank (the shallowest tank) having a specific $K_L a$ higher than the baseline value, but the error is negligibly small. Although the rating curves in Fig. 3.9 show that the $(K_L a)_{20}$ values are always less than the baseline $(K_L a_0)_{20}$, it is generally accepted that the deeper the tank, the higher the oxygen transfer efficiency, all things else being equal [Houck and Boon 1980, Yunt and Hancuff 1988a, 1988b] [EPA/625/1-89/023 (1989)]. This is simply because the dissolved oxygen saturation concentration increases with depth, which offsets the loss in the transfer coefficient in a deep tank. The net result is therefore still an increase in the overall aeration efficiency. Other clean water studies showed a nearly linear correlation between oxygen transfer efficiency and depth up to at least 6.1 m (20 ft) [Houck and Boon 1980]. The rating curves show that, in general, $K_L a$ decreases with depth at a fixed average volumetric gas flowrate. For the gas flowrate of 3.3 m^3/min, for example, the profile is almost linear up to 6 m, which confirms Downing and Boon's finding [Boon 1979] as mentioned in the Section 3.1 for the model development.

It is interesting to observe that, for deeper tank depths, the trend is not always decreasing, but actually starts to increase beyond a certain depth. This should be confirmed by further exploratory testing.

The exponential functional relationship between the mass transfer coefficient and tank depth presented herewith may explain this previously inexplicable phenomenon.

The predicted standard oxygen transfer efficiency (SOTE) using the simulation model (Eq. 3.6) can be compared with the actual measured SOTE based on the reported values (Yunt and Hancuff 1988a). Figure 3.10 below shows the compared results plotted in ascending order of the tank depths. Within experimental errors and simulation errors, the results seem to match very well.

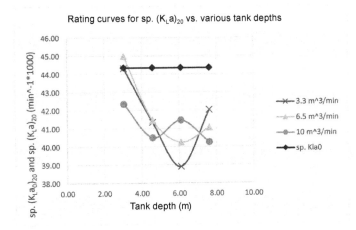

FIGURE 3.9 Rating curves for the standard specific transfer coefficients (K_La_0 and $K_La)_{20}$ for various tank depths and air flow rates

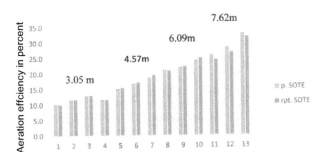

FIGURE 3.10 Comparison of the aeration efficiencies simulated vs. actual data

It would appear from the figure that the oxygen transfer efficiency is an increasing function of depth, even though the gas flow rates were not exactly the same for all the tests.

3.6 POTENTIAL FOR FUTURE APPLICATIONS

3.6.1 Scaling Up

In the application for scaling up, a clean water test must be performed. Most clean water testing is performed by using the clean water standard developed by the American Society of Civil Engineers (ASCE/EWRI 2-06). This standard uses a procedure that requires the test water (tap water) to be deoxygenated and then reaerated with the test diffusers at the appropriate airflow rate. The Standard was created for full-scale testing, not for small-scale testing to serve the purpose of scaling up for a project. The Standard urges that similar geometries be used for testing and design; however, the tank depth can be fixed at 3 m (10 feet) or 5 m (15 feet) or any other depth of choice.

Other differences exist because of the smaller scale. The data should be analyzed in strict adherence to the Standard; the non-linear estimation procedure was used and the time to complete the test ($\sim 4/K_La$ or 98% of equilibrium) will always be followed. It is preferable to repeat each test several times to have a constant K_La, and the test is to be repeated for different applied gas flow rates (average gas flow rate can be calculated from the standard gas flow rate), so that the K_La vs. Q_a relationship can be estimated. Once the baseline K_La is established, Eq. (3.6) can be applied to find the transfer coefficient at another tank depth. The effect of tank depth on the OTE is not just due to gas depletion, but also due to the natural volumetric expansion of the bubbles as they rise to the surface. To solve these complex phenomena, an Excel spreadsheet using the built-in software Solver is used to solve the simultaneous equations, using the established baseline parameter K_La_0, as well as the actual environmental conditions surrounding the second tank (see Table 3.3, but this time K_La_0 is a data, and the K_La becomes a variable to be determined). In other words, the same spreadsheet calculation method is used twice to calculate both K_La_0 and K_La. The proposed general procedure for estimating the specific baseline and the standard specific baseline $(K_La_0)_{20}$ is shown in the flow chart (Fig. 5.1) in Chapter 5.

3.6.2 Translation to In-process Oxygen Transfer

In the application for wastewater treatment, using the transfer of oxygen to clean water as the datum, it may then be possible to determine the equivalent bench-scale oxygen transfer coefficient $(K_La_{f_0})$ for a wastewater system, and the ratio of the two coefficients can then be used as a correction factor to be applied to fluidized systems treating wastewaters via aerobic biological oxidation, where microbial respiration has a significantly different contribution to gas depletion compared to clean water. However, before any mass balance equations can be used to evaluate this difference in the gas depletion rates, it is paramount to determine alpha (α) where alpha is the correction factor (Rosso and Stenstrom 2006) given by:

$$\alpha = \frac{K_La_{f_0}}{K_La_0} \approx \frac{K_La_f}{K_La} \tag{3.16}$$

It is postulated that this correction factor (α) can be determined by bench scale experiments.

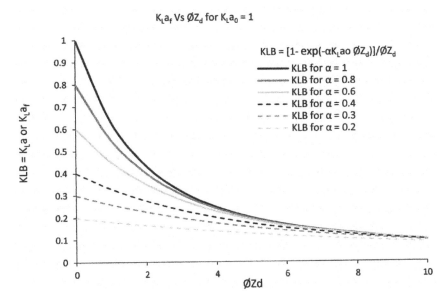

FIGURE 3.11 The apparent $K_L a_f$ plotted against $\emptyset Z_d$ for $K_L a_0 = 1$

It is hypothesized that this alpha value is not dependent on the liquid depth and geometry of the aeration basin and the model developed that relates $K_L a$ to depth then allows the alpha value to be used for any other depths and geometry of the aeration basin.

Therefore, using Eq. (3.6), after incorporating α into the mass transfer coefficient for in-process water, the mass transfer coefficient in in-process water $K_L a_f$ would be given by:

$$KLB = K_L a_f = \frac{1 - \exp(-\Phi Z_d \cdot \alpha\, K_L a_0)}{\Phi Z_d} \tag{3.17}$$

(where $KLB = K_L a$ for water ($\alpha = 1$) or $K_L a_f$ for wastewater ($\alpha < 1$))

This equation can be plotted for $K_L a_f$ against the function $\emptyset Z_d$ for when the baseline is unity, for various α values, as shown in Fig. 3.11 above.

The use of this equation for in-process water parameter estimations will be the subject of another paper to be submitted, pending further investigations. However, this subject is discussed in great length in Chapter 6. This graph of Fig. 3.11 shows exactly what Boon (1979) has found in his experiments, that $K_L a_f$ is a declining trend with respect to increasing depth of the immersion vehicle of gas supply.

3.7 CONCLUSIONS

The objective of this paper is to introduce a baseline oxygen mass transfer coefficient ($K_L a_0$), a hypothetical parameter defined as the oxygen transfer rate coefficient at zero depth, and to develop new models relating $K_L a$ to the baseline $K_L a_0$ as a function of temperature, system characteristics (e.g. the gas flow rate, the diffuser depth Z_d), and the oxygen solubility (C_S). Results of this study indicate that a uniform value of K_La_0 that is independent of tank depth can be obtained experimentally. This new mass transfer coefficient, K_La_0, is introduced for the first time in literature and is defined as the baseline volumetric transfer coefficient to signify a baseline. This baseline, K_La_0, has proven to be universal for tanks of any depth when normalized to the same test conditions, including the gas flow rate U_g (commonly known as the superficial velocity when the surface tank area is constant). The baseline K_La_0 can be determined by simple means, such as a clean water test as stipulated in ASCE 2-06.

The developed equation relating the apparent volumetric transfer coefficient (K_La) to the baseline (K_La_0) is expressed by Eq. (3.6).

The standard baseline $(K_La_0)_{20}$, when normalized to the same gas flowrate, is a constant value regardless of tank depth. This baseline value can be expressed as a specific standard baseline when the relationship between $(K_La_0)_{20}$ and the average volumetric gas flow rate Q_{a20} is known. Therefore, the standard baseline $(K_La_0)_{20}$ determined from a single test tank is a valuable parameter that can be used to predict the $(K_La)_{20}$ value for any other tank depth and gas flowrate (or U_g (height-averaged superficial gas velocity)) by using Eq. (3.6) and the other developed equations, provided the tank horizontal cross-sectional area remains constant and uniform as the bubbles rise to the surface. The effective depth 'de' can be determined by solving a set of simultaneous equations using a spreadsheet Solver, but, in the absence of more complete data, 'e' can be assumed to be between 0.4 to 0.5 (Eckenfelder 1970).

Therefore, $(K_La_0)_{20}$ can be used to evaluate the K_La in a full-size aeration tank (e.g. an oxidation ditch with a closed loop flow condition) without having to measure or estimate numerically the bubble size needed to estimate the K_La for such simulation. However, the proposed method herewith may require multiple testing under various gas flowrates, and preferably with testing under various water depths as well, so that the model can be verified for a system. Using the baseline, a family of rating curves for $(K_La)_{20}$ (the standardized K_La at 20°C) can be constructed for various gas flow rates applied to various tank depths. The new model relating K_La to the baseline K_La_0 is an exponential function, and $(K_La_0)_T$ is found to be inversely proportional to the oxygen solubility $(C_S)_T$ in water to a high degree of correlation. Using a pre-determined baseline K_La_0, the new model predicts oxygen transfer coefficients $(K_La)_{20}$ for any tank depths to be within 1~3% error compared to observed measurements and similarly for the standard oxygen transfer efficiency (SOTE%).

Hopefully, the problem with energy wastage due to inaccurate supply of air is ameliorated and the current energy consumption practice could be improved by applying the models to estimate the mass transfer coefficient (K_La) correctly for different tank depths at the design stage. As a benefit this analysis appears to support the temperature correction model [Lee 2017] as shown by Fig. 3.7 (for the baseline, K_La_0) and Fig. 3.8 (for K_La), showing the excellent regression correlations when the baseline is used in conjunction with the temperature model. Using the baseline K_La_0 is tantamount to using a shallow tank, which is the fundamental basis for the 5th power temperature model.

Although the mass transfer of oxygen in clean water is well researched and documented in the literature, its application in wastewater process conditions is not well understood. The development of Eq. [3.16] and Eq. [3.17] may lead to better relationships between clean water and process water in anattempt to elucidate the alpha factor (α) which currently appears to be a complicated function of process variables. The discovery of a standard baseline $(K_L a_0)_{20}$ that may be determined from shop tests for predicting the $(K_L a)_{20}$ value for any other aeration tank depth and gas flowrate, and even for in-process water with an alpha (α) factor incorporated into the equation as Eq. [3.17], is important. This finding may be utilized in the development of energy consumption optimization strategies for wastewater treatment plants. This work may also improve the accuracy of aeration models used for aeration system evaluations.

REFERENCES

ASCE-2-06. (2007). Measurement of Oxygen Transfer in Clean Water. Standards ASCE/EWRI. ISBN-10: 0-7844-0848-3, TD458.M42 2007.

ASCE-18-96. (1997). Standard Guidelines for In-Process Oxygen Transfer Testing. ASCE Standard. ISBN-0-7844-0114-4, TD758.S73.

Baillod, C.R. (1979). Review of oxygen transfer model refinements and data interpretation. Proc., Workshop toward an Oxygen Transfer Standard, U.S. EPA/600-9-78-021, W.C. Boyle, ed., U.S. EPA, Cincinnati, 17-26.

Boogerd, F.C., Bos, P., Kuenen, J.G., Heijnen, J.J., Van der Lans, R.G.J.M. (1990). Oxygen and carbon dioxide mass transfer and the aerobic, autotrophic cultivation of moderate and extreme thermophiles: a case study related to the microbial desulfurization of coal. Biotechnology and bioengineering 35(11): 1111-1119.

Boon, A.G. (1979). Oxygen Transfer in the Activated-Sludge Process. Water Research Centre, Stevenage Laboratory, England, United Kingdom.

Daniil, E.I., Gulliver, J.S. (1988). Temperature dependence of liquid film coefficient for gas transfer. Journal of Environmental Engineering Oct 1988, 114(5): 1224-1229.

Downing, A.A., Boon, A.G. (1968). Oxygen transfer in the activated sludge process. In: Eckenfelder, W.W. Jr., McCabe, B.J. (eds.). Advances in Biological Waste Treatment, MacMillian Co., NY, p. 131.

Eckenfelder, W.W. (1952). Aeration efficiency and design: I. measurement of oxygen transfer efficiency. Sewage and Industrial Wastes October 1952, pp. 1221-1228.

Eckenfelder, W.W. (1970). Water Pollution Control. Experimental Procedures for Process Design. The Pemberton Press, Jenkins Publishing Company, Austin and New York.

EPA/600/2-83-102. (1983). "Development of Standard Procedures for Evaluating Oxygen Transfer Devices" Municipal Environmental Research Laboratory Office of Research and Development, US Environmental Protection Agency Cincinnati, OH 45268.

EPA/600/S2-88/022. (1988). Project Summary – Aeration Equipment Evaluation: Phase I – Clean Water Test Results. Water Engineering Research Laboratory Cincinnati OH 45268.

EPA/625/1-89/023. (1989). "Fine pore aeration systems", U.S. Environmental Protection Agency, Office of Research and Development, Center for Environmental Research Information: Risk Reduction Engineering Laboratory.

Fayolle, Y., Cockx, A., Gillot, S., Roustan, M., Héduit, A. (2007). Oxygen transfer prediction in aeration tanks using CFD. Chemical Engineering Science 62(24): 7163-7171.

Houck, D.H., Boon, A.G. (1980). Survey and Evaluation of Fine Bubble Dome Diffuser Aeration Equipment, EPA/MERL Grant No. R806990, September, 1980.

Hunter III, J.S. (1979). Accounting for the effects of water temperature in aerator test procedures. Proceedings: Workshop Toward an Oxygen Transfer Standard, EPA-600/9-78 (Vol. 21, pp. 85-9).

Hwang, H.J., Stenstrom, M.K. (1985). Evaluation of fine-bubble alpha factors in near full-scale equipment. Journal WPCF, Volume 57, Number 12, U.S.A.

Jackson, M.L., Shen, C-C. 1978. Aeration and mixing in deep tank fermentation systems. AIChE Journal 24(1): 63.

Jiang, P., Stenstrom, M.K. (2012). Oxygen transfer parameter estimation: impact of methodology. Journal of Environmental Engineering 138(2): 137-142 · February 2012. DOI: 10.1061/(ASCE)EE.1943-7870.0000456.

Lakin, M.B., Salzman, R.N. (1977). Subsurface Aeration Evaluation. Paper presented at the 50th Annual Conference, Water Pollution Control Federation, Philadelphia, 1977.

Lee, J. (1978). Interpretation of Non-steady State Submerged Bubble Oxygen Transfer Data. Independent study report in partial fulfillment of the requirements for the degree of Master of Science (Civil and Environmental Engineering) at the University of Wisconsin, 1978 [Unpublished].

Lee, J. (2017). Development of a model to determine mass transfer coefficient and oxygen solubility in bioreactors, Heliyon, Volume 3, Issue 2, February 2017, e00248, ISSN 2405-8440. http://doi.org/10.1016/j.heliyon.2017.e00248.

McGinnis, D.F., Little, J.C. (2002). Predicting diffused-bubble oxygen transfer rate using the discrete-bubble model. Water Research 36(18): 4627-4635.

Metcalf & Eddy, Inc. second edition. (1985). Wastewater Engineering: Treatment & Disposal. ISBN 0-07-041677-X.

Metzger, I. (1968). Effects of temperature on stream aeration. Journal of the Sanitary Engineering Division 94(6): 1153-1160.

Stenstrom, Michael K. (2001). Oxygen transfer report: clean water testing (in accordance with latest ASCE standards) for Air Diffusion Systems submerged fine bubble diffusers on March 7, 8, 9, & 10 in 2001. http://www.aqua-sierra.com/wp-content/uploads/ads-full-oxygen-report.pdf.

Rosso, Diego, Stenstrom, Michael K. (2006). Alpha Factors in Full-scale Wastewater Aeration Systems. 2006 Water Environment Foundation.

Vogelaar, J.C.T., KLapwijk, A., Van Lier, J.B., Rulkens, W.H. (2000). Temperature effects on the oxygen transfer rate between 20 and 55 C. Water Research 34(3): 1037-1041.

Yunt, F., Hancuff, T. (1988a). Aeration Equipment Evaluation-Phase 1 Clean Water Test Results. Los Angeles County Sanitation Districts, Los Angeles, California 90607. Municipal Environmental Research Laboratory Office of Research and Development. USEPA, Cincinnati, Ohio 45268.

Yunt, Fred W., Hancuff, Tim O. (1988b). EPA/600/S2-88/022 (1988). Project Summary – Aeration Equipment Evaluation: Phase I – Clean Water Test Results. Water Engineering Research Laboratory Cincinnati OH 45268.

Zhou, Xiaohong, Shi, Hanchang, Song, Yanqing, Wu, Yuanyuan. (2012). Evaluation of oxygen transfer parameters of fine-bubble aeration system in plug flow aeration tank of wastewater treatment plant. Journal of Environmental Sciences 2013, 25(2) ISSN 1001-0742 CN.

4

The Lee-Baillod Equation

4.0 INTRODUCTION TO DERIVATION OF THE LEE-BAILLOD MODEL

Conceptually, before reaching the saturation state in a non-steady state test, since the oxygen concentration in water is less than would be dictated by the oxygen content of the bubble, Le Chatelier's principle requires that the process in the context of a bubble containing oxygen and rising through water with a dissolved-oxygen deficit, relative to the composition of the bubble, would seek an equilibrium via the net transfer of oxygen from the bubble to the water [Mott 2013]. In this scenario, even for the ultimate steady-state, oxygen goes in and out of the gas stream depending on position and time of the bubble of the unsteady state test. In clean water, one can view the mass balances as having two sinks-one by diffusion into water, and the other by diffusion from water back to the gas stream which serves as the other sink. Whichever is the greater depends on the driving force one way or the other. At system equilibrium, these two rates are the same at the equilibrium point of the bulk liquid, the equilibrium point being defined by 'de' in ASCE 2-06 [ASCE 2006]. At steady state, the entire system is then at a *dynamic* equilibrium, with gas depletion at the lower half of the tank below the 'de' level, and gas absorption back to the gas phase above de, the two movements balancing each other out. The expression applicable to a stream of gas bubbles undergoing gas transfer in a tank is given by Eq. [4.1]. The standard mass transfer equation (the Standard Model) is usually written as:

$$\frac{dC}{dt} = K_L a (C_\infty^* - C) \tag{4.1}$$

where $K_L a$ is defined in the ASCE 2-06 standard as the apparent volumetric mass transfer coefficient; C_∞^* is the determination point of the steady-state DO (dissolved oxygen) saturation concentration as time approaches infinity. This standard model can be derived by using the principle of conservation of mass, when C is the dissolved oxygen concentration (mg/L) at time t (min). The mathematical method employs the concept of substantial derivative or the "derivative following the motion", where the transfer is being observed as the bubble ascends to the surface. In considering an oxygen balance on a rising bubble, the transfer rate is given by the oxygen flux integrated over the bubble surface.

4.1 DERIVATION OF THE CONSTANT BUBBLE VOLUME MODEL

4.1.1 Conservation of Mass in the Gas Phase

Hence, the total rate of mass increase is related to the transfer rate by:

$$\frac{dM}{dt} = -\int_A \overline{J \cdot dA} \qquad (4.2)$$

where M is the mass of oxygen inside the bubble. J is the flux. The bar indicates it is a vector quantity. 'A' is the overall interfacial area of the bubble.

The flux N is given by

$$J = -K_G(P^* - P_G) \qquad (4.3)$$

where P^* is the saturation gas content corresponding to the dissolved oxygen concentration C and P_G is the partial pressure of oxygen in the gas phase; K_G is the overall mass transfer coefficient in the gas phase.

The flux equation can be expressed based on Dalton's Law as:

$$J = -K_G(C/H - yP) \qquad (4.4)$$

where H is Henry's Law constant and y is the oxygen mole fraction and P is the total pressure.

Therefore, the mass balance equation (Eq. 4.2) becomes

$$\frac{dM}{dt} = -\int_A K_G\left(yP - \frac{C}{H}\right)dA \qquad (4.5)$$

where M is given by the universal gas law as:

$$M = \frac{yPV_B}{RT} \qquad (4.6)$$

where V_B is bubble volume and R is specific gas constant of oxygen.

Applying the concept of substantial derivative, and assuming the flux to be constant over the bubble area, we have,

$$\frac{dM}{dt} = \frac{\partial}{\partial t}(M) + vb\frac{\partial}{\partial z}(M) = -K_G\left(yP - \frac{C}{H}\right)(a'V_B) \qquad (4.7)$$

where z is the vertical ordinate of the bubble in the tank or the distance through which the bubble has traveled; a' is the interfacial area per unit bubble volume, and is assumed constant; and vb is the velocity of bubble, which can be assumed constant when the radius of the bubble falls within the following range [McGinnis and Little 2002]:

$$7.0 \times 10^{-4} < r < 5.1 \times 10^{-3} \qquad (4.8)$$

where r is the radius of bubble in meters. Conceptually, the first term on the right in Eq. (4.7) represents the local time rate of change of mass. This can be neglected because the gas flow rate is rapid in the liquid column, so that the oxygen content in the gas phase instantaneously present in the aeration liquid column is small, compared to the total amount of oxygen passing through the liquid column. In other words, if the bulk aqueous-phase concentration does not change significantly during the time a bubble takes to rise through the tank, the pseudo-steady-state assumption may be invoked. The overall gas phase mass transfer coefficient is related to the individual mass transfer coefficients as

$$1/K_G = 1/k_G + 1/H\,(1/k_L) \tag{4.9}$$

where k_G and k_L are mass transfer coefficients for the gas film and liquid film, respectively, in accordance with the two-film theory [Lewis and Whitman 1924].

Note that for a very soluble gas, H is large and the second term becomes negligible so that

$$\frac{1}{K_G} \approx \frac{1}{k_G} \tag{4.10}$$

Similarly, the overall liquid phase mass transfer coefficient can be related to the individual mass transfer coefficients as follows:

$$\frac{1}{K_L} = \frac{1}{k_L} + H\left(\frac{1}{k_G}\right) \tag{4.11}$$

For a slightly soluble gas such as oxygen, H is very small and therefore the second term on the right-hand side of the equation becomes negligible, so that

$$1/K_L \approx 1/k_L \tag{4.12}$$

When the liquid film controls, $k_G >>> k_L$, so that

$$K_G \approx H \cdot K_L \tag{4.13}$$

Neglecting the first term and substituting (Eq. 4.13) into (Eq. 4.7), the equation (Eq. 4.7) reduces to:

$$vb \cdot \frac{d}{dz}(M) = -HK_L\left(yP - \frac{C}{H}\right)(a'V_B) \tag{4.14}$$

Since $a' = 6/d$, where d is the diameter of bubble, and

$$V_B = 4/3\pi r^3$$

where r is the radius of bubble, therefore,

$$vb\frac{d}{dz}(M) = -K_L(HyP - C)(4\pi r^2) \tag{4.15}$$

If M is expressed as the molar flow rate of gas instead of mass, and N is the number flux of bubbles entering the tank, the above equation can be written as

$$\frac{d}{dz}(M') = -K_L(HyP - C)\frac{(4\pi r^2)N}{vb} \tag{4.16}$$

where M' is the molar gas flow rate of oxygen.

McGinnis and Little (2002) derived a similar equation [McGinnis and Little 2002].

The relationship between M and M' is therefore given by:

$$M' = dM/dt \tag{4.17}$$

where N is given by

$$N = Q_0/V_0 \tag{4.18}$$

where V_0 is the initial bubble volume and Q_0 is the actual volumetric gas flow rate at the diffuser.

Therefore,

$$M_0' = y_0 P_0 \frac{Q_0}{RT_0} \tag{4.19}$$

where M_0' is the initial molar gas flow rate of gaseous oxygen or nitrogen; y_0 is the initial mole fraction of the gas, P_0 is the standard pressure, Q_0 is the gas flow rate at standard temperature and pressure (0°C and 1 bar), R is the ideal gas constant for oxygen, and T_0 is the standard temperature.

When M' is a variable,

$$M' = yPQ/(RT) \tag{4.20}$$

$$\left(\frac{1}{RT}\right) \cdot \frac{d}{dz}(yPQ) = -K_L \frac{(HyP - C)(4\pi r^2)N}{vb} \tag{4.21}$$

Assuming a uniform cross-sectional area of the tank S, then

$$S \cdot \frac{U_g}{RT} \cdot \frac{d(yP)}{dz} = -\frac{4\pi r^2 N}{vb} \cdot K_L \cdot H \cdot P\left(y - \frac{C}{HP}\right) \tag{4.22}$$

where U_g is the average gas superficial velocity over the tank cross-sectional area S given by $U_g = Q_a/S$. Here, an important assumption is made: that the bubble volume remains constant as it rises to the surface, so that r is constant. This assumption is made because of the limitations of the state of the art of solving calculus, without which the differential equation Eq. (4.22) cannot be solved. [McGinnis and Little [2002] use numerical integration to Eq. (4.22) for both oxygen and nitrogen, in order to obtain the change in the molar flow rate, while the gas bubble is in contact with the water in the aeration tank. They assumed that vb and K_L are functions of r]. However, as mentioned before, vb can be assumed constant for a certain range of r, but the assumption that r is constant with respect to depth is more difficult to justify. This is because the bubble radius increases in response to decreasing hydrostatic pressure as well as the amount of oxygen and nitrogen transferred between the bubble and the water. While the latter gas-exchange effect can be ignored because of the pseudo-steady-state assumption, the first effect from the changes in hydrostatic pressure cannot be ignored because of Boyle's Law. The assumption of a constant r is therefore purely for facilitating the solution of the differential equation to solve the equation mechanistically instead of by numerical integration which required that the incremental results in the changes of partial pressure of oxygen and nitrogen within the bubble to be recalculated at incremental steps as the bubble rises through the tank. However, this assumption of r is applicable to a stream of gas bubbles of approximately equal bubble volumes in a shallow tank.

4.1.1.1   First Order Linear Differential Equation

The mathematical derivation of the Constant Bubble Volume model is based on the first order type of a differential equation,

$$dy/dx + f(x) \cdot y = g(x) \qquad \text{(i)}$$

First, let $f(x) = a$, where a is a given constant, so that

$$dy/dx + ay = g(x) \qquad \text{(ii)}$$

The left side is not an exact differential but can be made so by multiplying by an integrating factor. If $g(x) = 0$, equation (ii) has the solution $y = Ae^{-ax}$, or $ye^{ax} = A$, i.e. $d(ye^{ax}) = 0$. This suggests that the left side of equation (ii) can be made an exact differential by multiplication by e^{ax}.

Multiplying both sides of equation (ii) by e^{ax} ($\neq 0$) gives

$$e^{ax}dy/dx + ae^{ax}y = e^{ax}g(x)$$

or

$$d/dx(ye^{ax}) = e^{ax}g(x)$$

Integrating,

$$ye^{ax} = \int e^{ax}g(x)\,dx + k$$

or

$$y = e^{-ax}\int e^{ax}g(x)\,dx + ke^{-ax} \qquad \text{(iii)}$$

where k is an arbitrary constant.

Next, consider the more general equation (i). This can be made exact by multiplication by an integrating factor $q(x)$ to be determined. Let $f(x) = dq/dx$, or $q = \int f(x)\,dx$, then equation (i) becomes

$$dy/dx + dq/dx\, y = g(x)$$

Multiplication by e^q gives

$$d/dx(e^q y) = e^q g(x),$$

from which, on integration,

$$e^q y = \int e^q g(x)\,dx + k' \qquad \text{(iv)}$$

where k' is an arbitrary constant.

Since

$$q = \int f(x)\,dx$$

is known, y is given in terms of x. Hence, multiplying equation (i) by the integrating factor $\exp(\int f(x)\,dx)$ makes the left side of equation (i) exactly integrable. Therefore, from (Eq. 4.22),

$$S \cdot \frac{U_g}{RT} \cdot \frac{d(yP)}{dz} = -\frac{4\pi r^2 N}{vb} \cdot K_L \cdot H \cdot P\left(y - \frac{C}{HP}\right) \tag{4.23}$$

Rearranging,

$$\frac{d(yP)}{dz} = -\frac{4\pi r^2 N}{v_b \cdot U_g} \cdot K_L \cdot H \frac{RT}{S} \cdot P\left(y - \frac{C}{HP}\right) \tag{4.24}$$

letting

$$\Omega = \frac{4\pi r^2 N}{v_b \cdot U_g} \cdot K_L \cdot H \frac{RT}{S} \tag{4.25}$$

Therefore,

$$\frac{d(yP)}{dz} + \Omega P y = \frac{\Omega C}{H} \tag{4.26}$$

Since both y and P are functions of z, the equation can be expanded as,

$$\frac{ydP}{dz} + \frac{Pdy}{dz} + \Omega P y = \frac{\Omega C}{H} \tag{4.27}$$

Hence,

$$\frac{dy}{dz} + y\left(\Omega + \frac{d(\ln P)}{dz}\right) = \frac{\Omega C}{HP} \tag{4.28}$$

Assuming vb to be constant, this is a first order linear differential equation with non-constant coefficients just like equation (iv), and letting

$$q = \int\left(\Omega + \frac{d(\ln P)}{dz}\right)dz \tag{4.29}$$

integrating gives

$$q = \Omega z + \ln P \tag{4.30}$$

Letting $x = z$ and $g(x) = \Omega C/(HP)$, equation (iv) becomes

$$y = \exp(-(\Omega z + \ln P))\left[\int \frac{\Omega C}{HP} \cdot \exp(\Omega z + \ln P) \cdot dz + k'\right] \tag{4.31}$$

Simplifying,

$$yP \exp(\Omega z) = \frac{C}{H}\exp(\Omega z) + k' \tag{4.32}$$

where k' is an integration constant.

The boundary condition is that at $z = 0$, $y = Y_0$, and $P = P_d$ with the datum of origin at the level of the diffuser orifice. Y_0 is the initial oxygen mole fraction at the diffuser depth usually assumed to be 0.21; P_d is the hydrostatic pressure at the diffuser depth. Using the boundary values, the definite integral becomes:

$$y = \frac{C}{HP} + \left(\frac{Y_0 P_d}{P} - \frac{C}{HP} \right) \exp(-\Omega z) \tag{4.33}$$

The above equation represents the oxygen mole fraction at any depth z measured from the bottom. When y is plotted against z, an oxygen mole fraction variation curve is obtained. In the CBVM (Constant Bubble Volume Model), this mole fraction variation curve is a special case, where the inert gas in the bubble is continually being vented in such a way that the expansion of the bubble, as it rises to the surface, is compensated by the loss of the inert gas, so that the bubble volume remains constant. The derivative dy/dz in this equation is always positive, indicating that the equation is an ever-increasing function and it makes sense, since the ever depletion of the inert gas must correspond to an ever-increasing oxygen mole fraction. For the general case where the inert gas mole fraction is constant within the bubble, the equation can be adjusted by calibration factors introduced to the two 'independent' variables P and z.

4.1.2 Conservation of Mass in the Liquid Phase

For the bulk liquid, the net accumulation rate of dissolved gas in the liquid column is equal to the gas mass flow rate delivered by diffusion, if there is no gas escape from the liquid column. (This assumption is not entirely satisfactory because the liquid column may be supersaturated with the dissolved gas with respect to any gas phase outside the liquid column as C approaches C_∞. In the case of air aeration, dissolved oxygen from the bulk liquid will flow out to the atmosphere in addition to the gas stream exit). The diffusional flux is again given by the two-film theory as:

$$N = K_L (C^* - C),$$

where C^* is the dissolved gas saturation concentration in the liquid phase; K_L is the overall mass transfer coefficient in the liquid phase.

or,

$$\frac{dw}{dt} = \int K_L a (HyP - C) dV \tag{4.34}$$

but since $w = CV$, the differential equation becomes

$$\frac{dC}{dt} = \frac{K_L a}{Z_d} \int_0^{Z_d} (HyP - C) dz \tag{4.35}$$

integral being applied from 0 to Z_d, where Z_d is the total depth.
Therefore,

$$C = \frac{K_L a}{Z_d} \int_0^t \int_0^{Z_d} (HyP - C) dz dt \tag{4.36}$$

Substituting y in Eq. (4.33) into Eq. (4.36) as derived previously, the first integral with respect to z can be solved by integration, recognizing that P is also a function of z, resulting in the following:

$$\int_0^{Zd} (HyP - C)\,dz = \frac{(HY_0P_d - C)(1 - \exp(-\Omega Z_d))}{\Omega Z_d} \tag{4.37}$$

Therefore, substituting Eq. (4.37) into Eq. (4.36), we have

$$C = K_L a \int_0^t (HY_0 P_d - C)\,dt \cdot \frac{(1 - \exp(-\Omega Z_d))}{\Omega Z_d} \tag{4.38}$$

As explained before, this equation is valid only when the bubble size is constant with depth and time. In other words, $K_L a$ was assumed to be constant with depth and time. In reality, the oxygen transfer film is affected by several factors, notably changes in pressure, and gas depletion, both of which are functions of depth and time. *Taking the parameter $K_L a$ out of the integral in Eq. 4.35 is an approximate mathematical treatment only.* However, the equation Eq. (4.38) would be approximately true if the aeration tank is shallow, so that the changes in hydrostatic pressure are small. Defining a baseline $K_L a$ as:

$$\lim_{Zd \to 0} K_L a = K_L a_0 \tag{4.39}$$

The equation can be written as

$$C = K_L a_0 \frac{(1 - \exp(-\Omega Z_d))}{\Omega Z_d} \int_0^t (HY_0 P_d - C)\,dt \tag{4.40}$$

In the differential form, the equation would become

$$\frac{dC}{dt} = K_L a_0 \frac{(1 - \exp(-\Omega Z_d))}{\Omega Z_d}(HY_0 P_d - C) \tag{4.41}$$

Since in a non-steady state clean water test, the standard model is as stated by Eq. 4.1,

$$\frac{dC}{dt} = K_L a (C_\infty^* - C) \tag{4.42}$$

Comparing Eq. (4.41) and Eq. (4.42), it is obvious that the two equations would match if

$$C_\infty^* = HY_0 \cdot P_d \tag{4.43}$$

$$K_L a = K_L a_0 \frac{(1 - \exp(-\Omega Z_d))}{\Omega Z_d} \tag{4.44}$$

The above set of equations (Eq. 4.43 and Eq. 4.44) represents the CBVM (Constant Bubble Volume Model). The model can be improved if it is recognized that the equilibrium concentration C_∞^* does not saturate at the bottom of the tank at the diffuser depth. Both the equilibrium pressure and the equilibrium mole fraction of oxygen are different from Eq. (4.43) and so a calibration is required. Furthermore, Eq. (4.43) can be written as,

$$C_\infty^* = HY_0 \cdot (\rho_w \cdot g \cdot Z_d + P_b - P_{vt}) \qquad (4.45)$$

where ρ_w is the density of water; g is the gravitational constant, and P_b is the barometric pressure and P_{vt} is the vapor pressure at the free surface.

4.1.3 Derivation of the Depth Correction Model

The CBVM recognized that both parameters $(K_L a)$ and (C_∞^*) are functions of Z_d. Comparing Eq. (4.41) and Eq. (4.42), since they are one and the same equation, it is easy to see that if it is required to apply a correction factor similar to what Downing and Boon [Downing and Boon 1968] did to their equation for C_∞^* to match with reality, then a corresponding correction for $K_L a$ is similarly required, and the common linkage between these two equations must be the submergence depth Z_d.

The equation for C_∞^* (Eq. (4.45)) appears to suggest that the saturation equilibrium level has occurred at the submergence depth Z_d. As the submergence depth is not the equilibrium level, which should occur at an effective depth d_e, an adjustment must be made in order to correctly predict the C_∞^* value. Since C_∞^* and $K_L a$ are co-related (approximately inversely proportional to each other), an adjustment to C_∞^* must have a corresponding adjustment to $K_L a$. Bearing in mind that C_∞^* is measured from the surface downward toward the bottom, whereas $K_L a$ was derived based on the travel distance of bubble measured from the bottom to the top, the point of origin of the parameters' individual reference frame is not the same but opposite to each other. If the adjustment to C_∞^* is $e \cdot Z_d$, the corresponding adjustment to the submergence depth for $K_L a$ must be $(1 - e)$, where e is the saturation depth correction for Eq. (4.45), given as

$$e = \frac{d_e}{Z_d} \qquad (4.46)$$

and P_b is the overburden atmospheric pressure on the liquid column. Therefore, (Eq. 4.45) becomes:

$$C_\infty^* = HY_0(\rho w g \cdot e Z_d + P_b - P_{vt}) \qquad (4.47)$$

This equation assumed that the oxygen mole fraction at the equilibrium level is the same as the initial mole fraction at the bottom Y_0, which is 0.21. In reality, as seen in Figure 3.1 in Chapter 3, the oxygen mole fraction at equilibrium is slightly less than Y_0 because of gas depletion prior to reaching this level, and so to compensate, the equilibrium mole fraction ye must be slightly smaller than y_0 of 0.21 at the true equilibrium level, as illustrated by Fig. 3.1. However, for the purpose of calibrating the model, this error in Y_e is often deemed acceptable and Eq. (4.47) is deemed to be valid [ASCE 2007]. *This equation has been used in the current ASCE Standards for determining 'de' based on clean water test results for C_∞^*.*

Another argument is that, since the correction factor 'e' is applied to Z_d for the calculation of C_∞^*, by mathematical induction or de facto implied by the analytic proof (Eq. 4.2 to Eq. 4.44) that the standard model (Eq. 4.1) is valid for the general case as well, the same correction factor is applied to Z_d in the mathematical equation for $K_L a$ because of the hypothesis that $K_L a_0$ and C_S are inversely proportional to each other [Lee 2017], but the correction factor is $(1 - e)$ because of the point of origin being fixed at the bottom. Therefore, the final model for $K_L a$ vs. $K_L a_0$ can be expressed as a variation of the special case of constant bubble equation (Eq. 4-44), as shown below by Eq. 4.48:

$$K_L a = K_L a_0 \frac{(1 - \exp(-\Omega(1 - e)Z_d))}{\Omega(1 - e)Z_d} \tag{4.48}$$

Ω given by Eq. (4.25) can be expressed in terms of $K_L a_0$ by recognizing that $a' = 6/d_b$, where d_b is the diameter of a spherical bubble (assumed constant), and $v_b = U_g(a'/a)$, as well as that

$$N = 3Q/4\pi r^3 \tag{4.48a}$$

Therefore,

$$\Omega = K_L a_0 \frac{HRST}{Q} \tag{4.49}$$

where a_0 is the interfacial area per unit of liquid volume V (at constant d_b) given by:

$$a_0 = \frac{6}{d_b} Q \frac{Z_d}{vb \cdot V} \tag{4.50}$$

where Q is the average gas flow rate of air or Q_a, V is the volume of the bulk liquid, and therefore Eq. (4.48) can be written as

$$K_L a = \frac{[1 - \exp(-K_L a_0 x(1 - e)Z_d)]}{x(1 - e)Z_d} \tag{4.51}$$

where

$$x = \frac{\Omega}{K_L a_0} \tag{4.52}$$

and is given by

$$x = HR \frac{ST}{Q} \tag{4.53}$$

where S is the horizontal cross-sectional area assumed to be uniform throughout the liquid column, and R is the specific gas constant for the oxygen gas under transfer. The parameter x is hereby defined as the *gas-flow constant*, when the average volumetric gas flow rate is fixed, for a specific temperature T.

The important assumptions for the derivation of this model are as follows:

- Only one gas is under transfer, all other gases inside bubbles are inert;
- Transferred gas is only slightly soluble so that the liquid film controls the transfer process;
- The liquid column within the confine of its boundary is well mixed so that the dissolved gas concentration is uniform;
- The bubbles assumed rising in a uniform column with uniform bubble size and velocity are corrected to an effective depth (*de*) by a depth ratio (*e*);
- Uniform horizontal cross-sectional area in the liquid column;

- The effects of gas hold-up, coalescence of bubbles or breaking up of bubbles, and gas transfer during the bubble formation stage are ignored. Any gas transfer at the surface of the liquid column has been ignored.

The effective saturation depth, de, represents the depth of water under which the total pressure (hydrostatic plus atmospheric) would produce a saturation concentration equal to C_∞^* for water in contact with air at 100% relative humidity. In a clean water test, it is calculated based on ASCE 2-06 (Section 8.1 and Annex F) [ASCE 2007]. The method given in the ASCE standard, however, is only an approximation.

Eq. (4.51) can also be written as:

$$K_L a = \frac{1 - \exp(-\phi Z_d \cdot K_L a_0)}{\phi Z_d} \quad (4.54)$$

where

$$\phi = x(1-e) \quad (4.55)$$

Eq. (4.54) is herewith defined as the Depth Correction Model. Since the parameter ϕ is an adjustment to the *gas-flow constant* x, it can be defined as *the effective gas-flow constant*. The compound parameter ϕZ_d is defined as the *characteristic depth* of the diffuser or the immersion depth of the aeration device. By rearranging Eq. (4.54), $K_L a_0$ can be obtained as:

$$K_L a_0 = -1/(\phi Z_d)\{\ln(1 - K_L a(\phi Z_d)\} \quad (4.56)$$

where $K_L a$ (the apparent $K_L a$) can be obtained from a curve fitting to clean water testing data using the ASCE 2-06 equation as shown here:

$$C = C_\infty^* - (C_\infty^* - C_0)\exp(-K_L at) \quad (4.57)$$

The implicit function (e) in Eq. (4.51) can be solved in a spreadsheet. Solving for (e) depends on solving the depth of equilibrium level (de). Eq. (4.54) and (4.55) constitute the proposed Depth Correction Model. It is envisaged that the baseline $K_L a_0$ as calculated by Eq. (4.56) will not change for any depth of tank under a fixed average volumetric gas flow rate (Q_a) for a system under testing.

4.1.4 The Hypothesis of a Constant Baseline ($K_L a_0$)

As mentioned before, this thesis advances the concept of a constant baseline for the mass transfer coefficient. For every tank tested using the non-steady state testing method, there is a baseline mass transfer coefficient that would be constant regardless of the tank height, as long as it was tested under the same average volumetric gas flow rate Q_a.

It would appear from the above derivation that the baseline mass transfer coefficient ($K_L a_0$) is indeed only dependent on the volumetric gas flow rate Q_a, although from (Eq. 4.44) the apparent mass transfer coefficient is not only a function of Q_a but also a function of depth Z_d. Based on the above calculations, the author advances the hypothesis that, for the same volumetric average gas flow rate, and the same water temperature and barometric pressure, the baseline mass transfer coefficient as calculated by (Eq. 4.44) would be a constant for any tank depth or diffuser depth

Z_d. In other words, a clean water test carried out on a 3.05 m (10 ft) tank would give the same $K_L a_0$ for the test carried out in a 4.57 m (15 ft) tank, or in a 6.09 m (20 ft) tank, or in a 7.62 m (25 ft) tank, as long as the volumetric gas flowrate is kept constant in all tests. A corollary of this finding is that, given a system of known depth of aeration, and given a supplied gas flow rate, it would be possible to estimate the mass transfer coefficient ($K_L a$) for any tank depth, without having to conduct a full-scale clean water test, if the baseline $K_L a_0$ is known. However, the determination of the baseline relies on an accurate determination of the equilibrium point of the saturation concentration, previously assumed to be at mid-depth [Eckenfelder 1970] but can be more accurately determined by the mole variation equation (Lee-Baillod model) over a tank height. The point at which the curve gives a minimum mole fraction is the equilibrium point. This can be determined by differentiation of the equation and at $dy/dx = 0$, the minimum point is obtained.

The model for the C_∞^* equation relies on the estimation of "e", for a correction term. This value of "e" estimated in one tank may not be used in a tank of another depth, since no test results have been presented to substantiate that "e" is more or less constant for tanks of different heights.

Even though the current ASCE clean water standard has mandated an effective-depth correction, this ASCE correction can only be applied to each individual tank under testing. This correction varies among different tank sizes, diffuser densities and layouts, and diffuser types, not to mention different tank depths. This chapter attempts to prove that, for all other conditions having been fixed, there is a unique relationship between $K_L a$ and tank depth, and that for as long as these conditions are fixed, the effective depth ratio (e) can be proven to be quite constant, as can be seen in Fig. 3.2 in the last chapter. Since a model has been developed based on the mole fraction variation along the tank height (Lee-Baillod model), this constant e value can be exploited for scale-up purposes as approximate solutions. Even though its usefulness in scaling up requires further study, the constant baseline $K_L a_0$ would be useful as a function of comparing aeration equipment tested under different tank depths, and for evaluation of performance for the purpose of rating aeration equipment. Furthermore, the baseline mass transfer coefficient can be used to determine the oxygen transfer coefficient in a wastewater aeration tank, using Eq. (3.17) given in Chapter 3 by incorporating the alpha factor (α) into the clean water depth correction model (Eq. 3.6), assuming that the in-process water mass transfer coefficient ($K_L a f_0$) can be determined on a bench scale. For in-process water, the model will depend on modifying the clean water gas transfer equations to include the additional gas depletion due to the microbial respiration as explained in Chapter 6. The application of alpha (α) so determined by bench-scale tests to determine the corresponding oxygen transfer coefficient in a full-scale situation is discussed in Chapter 5 and Chapter 6.

4.1.5   Determination of Calibration Parameters (n, m) for the Lee-Baillod Model

There are two methods to adjust the original equations. One method, by the correction factor 'e', has already been explained. For the other method, since the assumption of constant bubble volume and therefore constant interfacial area seriously compromises the Lee-Baillod model equation (Eq. 4.33) for the real case, the variables z and P need to be adjusted. Lee [1978] and Baillod [1979] postulated that the mole fraction variation curve is not linear, and certainly not an ever-increasing function as the CBVM (Constant Bubble Volume Model) has predicted. By assigning a parameter n to all the pressure values, and another parameter m to z, the equation can be written as:

$$y = \frac{C}{nHP} + \left(\frac{Y_0 n P_d}{nP} - \frac{C}{nHP}\right) \exp(-Hk \cdot mz) \tag{4.58}$$

where $Hk = \Omega$. Therefore,

$$k = K_L a_0 \, RST/Q_a \quad \text{or} \quad k = K_L a_0 \, RT/U_g$$

The above equation is hereby defined as the generalized Lee-Baillod Model and k can be defined as a *specific* baseline constant.

Rearranging the equation gives

$$yP = \frac{C}{nH} + \left(Y_0 P_d - \frac{C}{nH}\right) \exp(-Hk \cdot mz) \tag{4.59}$$

This equation allows the partial pressure of oxygen in the bubble at any location and at any time to be expressed. Rearranging Eq. (4.59) to solve for k as:

$$Hk = \frac{1}{mz}\left[\ln\left\{\frac{nHY_0 P_d - c}{n\,HYP - c}\right\}\right] \tag{4.60}$$

or,

$$K_L a_0 = \frac{Qa}{HRST}\Big/(mz)\left[\ln\left\{\frac{nHY_0 P_d - c}{n\,HYP - c}\right\}\right] \tag{4.61}$$

At the exit gas when $t = \infty$, $c = C_\infty^*$, Y (exit gas mole fraction) $= Y_0$, $P = P_a$ (the atmospheric pressure) and $Z = Z_d$, then

$$K_L a_0 = \left(\frac{1}{x}\right)\Big/(mZ_d)\left[\ln\left\{\frac{nHY_0 P_d - C_\infty^*}{nHY_0 P_a - C_\infty^*}\right\}\right] \tag{4.62}$$

where $x = HRST/Q_a$; $P_a = P_b - P_{vt}$

Rearranging Eq. (4.62) gives

$$C_\infty^* = nH \times Y_0 \times \frac{P_a - P_d \exp(-mx \cdot K_L a_0 \cdot Z_d)}{1 - \exp(-mx \cdot K_L a_0 \cdot Z_d)} \tag{4.63}$$

Eq. (4.58) is similar to Eq. (4.33) and after substituting y by Eq. (4.58) as derived previously into Eq. (4.36), the first integral with respect to z can be solved by integration w.r.t. z to give,

$$C_\infty^* = HY_0 P_d n \frac{1 - \exp(-mHkZ_d)}{1 - \exp(-mHkZ_d) + (n-1)mHkZ_d} \tag{4.64}$$

and,

$$K_L a = \frac{1 - \exp(-mx \cdot K_L a_0 \cdot Z_d)}{nmx \cdot Z_d} + \frac{(n-1)K_L a_0}{n} \tag{4.65}$$

In the application of the above equations, one equation (Eq. (4.64)) is not valid because of gas super-saturation at the free surface, where in the derivation any dissolved gas escaping into the atmosphere was ignored. This will tend to over-estimate C_∞^*, but the effect of supersaturation on the mass transfer coefficient is assumed to be small, and so Eq. (4.65) is considered valid, as this parameter is less sensitive to the DO concentration as C approaches C_∞^*. The equations that can be used to find the unknown parameters are:

1. oxygen mole fraction variations with depth: Eq. (4.58)
2. boundary conditions at the exit: Eq. (4.63)
3. integrated form of the mass transfer equation: Eq. (4.65)

Therefore, we have basically three equations to solve three unknowns, n, m, and $K_L a_0$. However, the solution does not include the effective depth de, which corresponds to the minimum mole fraction of the MF (mole fraction) curve at equilibrium. This can be achieved in two ways: plotting the MF curves using the MF equation (Eq. (4.58)) for a series of DO values upon knowing the n and m values, or, differentiating this equation and set dy/dz to zero to find the minimum point. In using the first method, a series of MF curves can be obtained as shown in the following example [EPA-600/2-83-102], where the tank height is 5.55 m (18.2 ft) and the saturation concentration was measured to be 11.43 mg/L as shown in Figure 4.1:

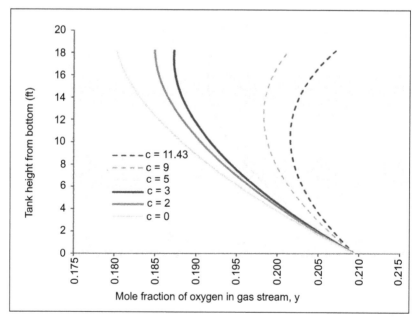

FIGURE 4.1 Off-gas mole fraction for a 5.55 m (18.2 ft) tank and the MF distribution curves

From the MF curve at saturation, the minimum mole fraction corresponding to the effective depth can be determined by graphical interpretation.

The parameters 'n' and 'm' can be considered as the characteristics of a particular aeration system at a fixed gas flow rate and can be determined by using a clean water test. The parameters would also serve as a correction for the underlying assumptions necessary for the model development, one of which is that any transfer of other gases from the bubbles, particularly nitrogen, was not incorporated. Bearing in mind that k is a function of the baseline $K_L a_0$, this equation (Eq. (4.58)) will eventually yield the result of the $K_L a_0$ in a clean water test as given by Eq. (4-62).

4.1.6   Determination of the Effective Depth 'd_e' or 'Z_e'

From Eq. (4.47)

$$C_\infty^* = HY_0(\rho w\, g \cdot eZ_d + P_b - P_{vt}) \qquad (4.66)$$

This equation assumed that the oxygen mole fraction at the equilibrium level is the same as the initial mole fraction at the bottom Y_0, which is 0.21. In reality, as seen in Chapter 3 Figure 3.1, the oxygen mole fraction at equilibrium is slightly less than Y_0 because of gas depletion prior to reaching this level, and so to compensate, ye must be slightly smaller than 0.21 to substitute for Y_0 in the above equation, at the true equilibrium level. This means e would be slightly bigger than the value calculated by (Eq. 4.66) when C_∞^* is known.

It was mentioned before that there are two methods to measure the system parameters, $K_L a$ and C_∞^*. Both methods would require the determination of the effective depth de. In EPA/625/1-89/023 as well as in ASCE 2-06 [ASCE 2007], de was determined based on Eq. (4.66) which assumed a constant mole fraction when steady state is reached. This assumption is an approximation that can be corrected by the following method with the following assumptions:

- mass balances of the gas absorption and desorption of a rising gas stream give a non-linear mole fraction curve (Eq. (4.58));
- the driving force is zero at this equilibrium level, so that the determination point C_∞^* is the saturation concentration at this level;
- The mole fraction at this level is at the minimum of the non-linear mole fraction variation curve which can be determined by differentiating Eq. (4.58) and setting it to zero.

Rigorous mathematical derivation showed that de is a function of environmental constants (barometric pressure, temperature, density of water), depth of tank, gas flux, $K_L a$, as well as C_∞^*, and, most importantly, the aeration system characteristics.

The following derived equation can be used to demonstrate to the European engineers who like to adopt a mid-depth correction that de is in fact a variable, and not necessarily mid-depth. This equation does not confirm that the saturation level is necessarily less than mid-depth-caution must be advised against jumping to the conclusion that saturation must occur less than mid-depth due to the inherent definition of de. The derivation of the equation to calculate the minimum mole fraction ye that corresponds to de is given below.

Recalling from Eq. (4.58) that the partial pressure of oxygen in a bubble is related to the depth of bubble measured as its distance from the diffuser depth, Z, the equation can be differentiated with respect to Z to find the minimum mole fraction, thus:

Eq. (4.58)

$$y = \frac{C}{nHP} + \left(\frac{Y_0 n P_d}{nP} - \frac{C}{nHP}\right)\exp(-Hk \cdot mz) \tag{4.67}$$

or

$$yP = \frac{C}{nH} + \left(Y_0 P_d - \frac{C}{nH}\right)\exp(-\Omega mz) \tag{4.68}$$

where $\Omega = Hk$, then

$$nH\,yP = C + ((nHY_0 P_d - C)\exp(-mxK_L a_0 z) \tag{4.69}$$

$$nH\left(y\frac{dP}{dz} + P\frac{dy}{dz}\right) = (nHY_0 P_d - C)(-mxK_L a)\exp(-mxK_L a_0 z) \tag{4.70}$$

For the boundary conditions at the minimum point:

$$dy/dz = 0 \text{ when } P = P_e,\, z = Z_e,\, y = Y_e,\, C = C_\infty^*$$

(where Y_e and P_e are the equilibrium mole fraction and the equilibrium total pressure, respectively).

Therefore,

$$Y_e = \frac{K_L a_0 mx}{nr_w H}(nHY_0 P_d - C_\infty^*)\exp(-mxK_L a_0 Z_e) \tag{4.71}$$

At equilibrium, according to Dalton's Law, the saturation concentration in the liquid phase C_∞^* is a function of the partial pressure in the gas phase given by:

$$C_\infty^* = HY_e P_e \tag{4.72}$$

where H is Henry's Law Constant. Substituting Y_e in Eq. (4.71) into Eq. (4.72), we have

$$C_\infty^* = \frac{mxK_L a_0}{nr_w}(nHY_0 P_d - C_\infty^*)\exp(-mxK_L a_0 Z_e)(P_e) \tag{4.73}$$

Rearranging terms, we have

$$Z_e = \frac{1}{mxK_L a_0}\left\{\ln\left(P_e\frac{mxK_L a_0}{nrw}\right) + \ln\left(\frac{nHY_0 P_d}{C_\infty^*} - 1\right)\right\} \tag{4.74}$$

where P_e is given by

$$P_e = P_a + rw(Z_d - Z_e) \tag{4.75}$$

where

P_a is the net pressure at the free surface given by:

$$P_a = P_b - P_{vt} \tag{4.76}$$

Ye can then be determined from Eq. (4.72) when C_∞^* is determined by a clean water test.

Eq. (4.74) gives one more equation to determine the five parameters, m, n, $K_L a_0$, Y_e and Z_e.

In the next chapter, examples are given to show how the $K_L a_0$ can be calculated for a set of clean water test result based on case studies of several experiments carried out by various investigators.

REFERENCES

ASCE-2-06. (2007). Measurement of Oxygen Transfer in Clean Water. Standards ASCE/ EWRI. ISBN-10: 0-7844-0848-3, TD458.M42 2007.

Baillod, C.R. (1979). Review of oxygen transfer model refinements and data interpretation. Proc., Workshop toward an Oxygen Transfer Standard, U.S. EPA/600-9-78-021, W.C. Boyle, ed., U.S. EPA, Cincinnati, 17-26.

Downing, A.A., Boon, A.G. (1968). Oxygen transfer in the activated sludge process. p. 131. In: Eckenfelder, W.W. Jr., McCabe, B.J. (eds.). Advances in Biological Waste Treatment, MacMillian Co., NY.

Eckenfelder, W.W. (1970). Water Pollution Control. Experimental Procedures for Process Design. The Pemberton Press, Jenkins Publishing Company, Austin and New York.

EPA/600/2-83-102. (1983). Development of Standard Procedures for Evaluating Oxygen Transfer Devices. Municipal Environmental Research Laboratory Office of Research and Development, US Environmental Protection Agency Cincinnati, OH 45268.

EPA/625/1-89/023. (1989). Fine pore aeration systems, U.S. Environmental Protection Agency, Office of Research and Development, Center for Environmental Research Information: Risk Reduction Engineering Laboratory.

Lee, J. (1978). Interpretation of Non-steady State Submerged Bubble Oxygen Transfer Data. Independent study report in partial fulfillment of the requirements for the degree of Master of Science (Civil and Environmental Engineering) at the University of Wisconsin, 1978 [Unpublished].

Lee, J. (2017). Development of a model to determine mass transfer coefficient and oxygen solubility in bioreactors, Heliyon, Volume 3, Issue 2, February 2017, e00248, ISSN 2405-8440, http://doi.org/10.1016/j.heliyon.2017.e00248.

Lewis, W.K., Whitman, W.G. (1924). Principles of Gas Absorption. Industrial and Engineering Chemistry 1924, 16(12), pp. 1215-1220 Publication Date: December 1924 (Article) DOI: 10.1021/ie50180a002.

McGinnis, D.F., Little, J.C. (2002). Predicting diffused-bubble oxygen transfer rate using the discrete-bubble model. Water Research 36(18): 4627-4635.

Mott, H.V. (2013). Environmental Process Analysis: Principles and Modeling. John Wiley & Sons. ISBN: 978-1-118-11501-5 December 2013.

5

Baseline Mass Transfer Coefficients and Interpretation of Non-steady State Submerged Bubble Oxygen Transfer Data

5.0 INTRODUCTION

The clean water standard 2-06 (ASCE 2007) was originally published to provide the industry with a tool that ensures all manufacturers provide data using the same methodology. The standard has been used successfully since it was first published in 1992. It has *recently* been revised and re-balloted through the consensus standards process to provide an updated standard, which is expected to be published soon. While variations in equipment in terms of individual performances are important, these have less adverse impacton parameter estimation than variations in the other factors, such as, in the past, variations in the test results obtained between test methods, as well as between different analyses of variance methods for the data. The emphasis on the non-linear least squares (NLLS) regression analysis method will have greatly re-assured manufacturers to the standard. The "Log Deficit" method, which requires *a priori* estimation of the equilibrium oxygen concentration (C_∞^*), is expected to be removed from the standard. This deletion will bring data interpretation and analysis to an even higher degree of accuracy and consistency [Jiang and Stenstrom 2012].

However, the effects of other variables, such as temperature, pressure, geometry, etc., still require deliberations in order to obtain reproducible results and more importantly, in the application of clean water results to field conditions which are subject to additional whole hosts of variables affecting oxygen transfer. Therefore, a systematic and progressive elimination of these effects is the way forward, and the proposals made in this paper appear to be the first step in this direction.

Although the clean water standard has fulfilled the purpose of setting a standard of conformance for manufacturers to use in the measurement of oxygen transfer, especially in the area of test methods and data interpretation, it falls short for the purpose of compliance testing, which is in fact the main purpose of the standard. The mass transfer coefficient ($K_L a$ or $K_L a_f$) is one of the most important parameters in the water and wastewater treatment technology. [ASCE 1997] [ASCE 2007] (The subscript f stands for any field obtained measurement). Testing in clean water eludes the many factors that affect the use of this parameter. One factor particularly for submerged diffused aeration, for example, is the varying gas-phase gas depletion rate. In in-process water, care must be taken to ensure that the parameter is *not* a function of dissolved oxygen concentration. This dependency can occur where air is blown through diffusers on the bottom of activated sludge tanks, where rising air bubbles are significantly depleted of oxygen as they ascend to the water surface. The extent of oxygen depletion is a function of the oxygen concentration in the activated sludge mixed liquor [Ahlert 1997] [Rosso and Stenstrom 2006].

The dissolved oxygen concentration C in a well-mixed tank is a function of the biological uptake rate R. Since the effect of this uptake rate on oxygen transfer is additive (negative in the context of the basic mass transfer model), the attendant gas depletion rate effect on the oxygen transfer must also be additive. This effect is not associative as a scaling quantity of $K_L a$ with the use of a scalar factor alpha (α), as is the current practice [Rosso and Stenstrom 2006]. Since the microbial gas depletion rate arising from microbial respiration comes only from the presence of microbes, the respiration rate R must equal the gas depletion rate, other minor factors impacting the transfer rate notwithstanding. Hence, in a batch process, in order to utilize the clean water mass transfer coefficient, the gas depletion rate (gdp) due to the microbes must be accounted for, that the author believes should result in an equation

$$\frac{dC}{dt} = K_L a_f (C^*_{\infty f} - C) - R - gdp$$

which, when equating R with gdp, gives:

$$\frac{dC}{dt} = K_L a_f (C^*_{\infty f} - C) - 2R$$

This is the only way to take into account the microbial gas depletion rate effect on the oxygen transfer in the in-process water. Simply multiplying an alpha factor associative to the mass transfer coefficient does not fully account for this effect. This is described in Chapter 6 and is the subject of another paper being published.

Similarly, gas depletion comes from other sources as well, such as tank height and its associated water depths. In a non-steady state test as described in the standard, the driving force as derived from the concentration gradient gradually diminishes as the dissolved oxygen increases until it reaches saturation, and so the gas-side gas depletion rate varies throughout the test [Baillod 1979] [DeMoyer et al. 2002] [McGinnis and Little 2002] [Schierholz et al. 2006] [Lee 2018]. Naturally, this spectrum of gas depletion would be dependent on the tank height, so that deeper tanks will have a higher overall gas depletion. Given that $K_L a$ is a function of many variables, including the water depth under test, in order to have a unified test result, it is necessary to create a baseline mass transfer coefficient, so that all tests will have the same measured baseline. A paper published by the author Lee [2018] examines the depth effect by the introduction of this baseline mass transfer coefficient ($K_L a_0$). The baseline would be *independent* of tank height, since it measures $K_L a$ for a tank of virtually zero height. Manufacturers who can calculate the baseline based on a series of testing and measurements with different gas velocities should be expected to produce a uniform constant value of the *specific baseline mass transfer coefficient* regardless of the tank heights they use. In this context, unlike the effect of microbes, gas depletion due to this source (tank height) is eliminated indirectly not so much by incorporating this effect into the transfer equation, but by changing the evaluation of a mass transfer coefficient to one of zero-depth tank. The previous paper has explained how this is done based on theoretical development and numerous testing of data reported in the literature widely available to the public. This has led to the discovery of several physical mathematical models in nature applicable to the calculation of gas-phase oxygen mass transfer in water for submerged bubble aeration [Lee 2017, 2018]. In this chapter, additional previously published aeration data by others were re-analyzed by conducting regression analyses to determine and to verify this concept of a *standardized specific baseline mass transfer coefficient* $(K_La_0)/Q_a^q$ so that it can be used to offer a standardized practice of measurement of oxygen transfer. Only when all the negative effects impacting oxygen transfer are eliminated can it be confidently proclaimed that the standard is successful, especially in terms of compliance testing. A baseline is invaluable in this regard.

5.1   THEORY

Reports on aeration equipment rely on the basic transfer equation. Of the two main parameters $(K_La$ and $C_\infty^*)$ pertaining to the standard transfer equation, changing variables affect both the equilibrium values for oxygen concentration and the rate at which transfer occurs. The former has been studied extensively over a range of variables, but similar work for the rate coefficient K_La is less abundant. The more relevant papers on diffused bubble aeration include McGinnis and Little (2002), whose discrete bubble model forms the basis for the models developed by the author [Lee 2018]. Other similar work includes McWhirter and Hutter (1989), which is the basis for subsequent development by DeMoyer et al. (2002) and Schierholz et al. (2006) that are now cited in this manuscript in the Discussion section. Works involving this parameter almost invariably focus on the oxygen transfer rate (OTR), which includes both the equilibrium concentration and the transfer coefficient together. Ashley et al. (2009) looked at the effect of air flow rate, depth of air injection, among other things, on the oxygen transfer. Graphs were plotted to show that K_La is a function of air flow rate, but each depth tested would give a different unique graph. There is no correlation between graphs of different depths because K_La would depend on the gas depletion rate which varies with different depths. While C_∞^* varies with depths as governed by Henry's Law, the physical law governing the way K_La varies with depth seems quite unknown. If the author is not mistaken, this is the first time a mechanistic model based on first principles was ever derived, and it is an exponential function given by Eq. 3.6, restated below as Eq. 5.2.

Analysis of bubble aeration depends on average values. Oxygen transfer rate depends on the average surface area of the bubbles and thus on the mean bubble diameter d_b. Eckenfelder [1966] used this to relate to the average gas flow rate [Schroeder 1977]. Since d_b depends on temperature and pressure, Q_a would require adjustment to temperature and pressure as well, otherwise the basic transfer model cannot be used correctly. The mass transfer coefficient K_La is dependent on the gas average volumetric flow rate (Q_a) passing through the liquid column [Hwang and Stenstrom 1985]. Q_a is expressed in terms of volume of gas per unit time and is calculated by the universal gas law, or Boyle's Law if the liquid temperature is uniform throughout the liquid column, and taking the arithmetic mean of the flow rates over the tank column. $K_L a$ is directly proportional to this averaged gas flow rate to power q, where q is usually less than unity (for fine bubble aeration) for water in a fixed column height and a fixed gas supply rate at standard conditions. This average gas flowrate Q_a is determined from the given gas flow at standard conditions Q_S (the subscript s represents standard). The salient equation to convert Q_S to Q_a is given by Lee (2017) and in Chapter 2, Eq. 2.25, and restated here as Eq. 5.1 below:

$$Q_a = 172.82 \times Q_S \times T_P \left[\frac{1}{P_P} + \frac{1}{P_b} \right] \tag{5.1}$$

where T_P is the gas temperature at the point of flow measurement, in Kelvin, assumed to be equal to the water temperature; P_p and P_b are the corresponding gas pressure and the barometric pressure, respectively (units are in Pa). This equation has assumed the standard air temperature to be 20°C or 293 K, and a standard air pressure of 1.00 atm (101.3 kPa).

The effect of changing depth on the transfer rate coefficient was explored by Yunt et al. [1988a], and, building on his research, the new depth-correction model relating $K_L a$ to the baseline $K_L a_0$ [Lee 2018] is expressed by:

$$K_L a = \frac{1 - \exp(-\Phi Z_d \cdot K_L a_0)}{\Phi Z_d} \tag{5.2}$$

where Φ is a constant dependent on the aeration system characteristics $x \cdot (1 - e)$ and Z_d is the immersion depth of the diffusers, where x and e are defined in Chapter 4 (Eq. 4.53 to Eq. 4.55) as: $x = HR_0 T/U_g$, where U_g is the height-averaged superficial gas velocity; R_0 is the specific gas constant for oxygen (note: a different symbol is used to distinguish it from the respiration rate R); T is the water temperature; e is the effective depth ratio $e = de/Z_d$. Therefore, $K_L a$ is an exponential function of this new coefficient $K_L a_0$ and their relationship is given by Eq. (5.2), where $K_L a$ is dependent on the height of the liquid column Z_d through which the gas flow stream passes. The error value of $K_L a$ was obtained by comparison of the numerical results from model solution and the experimental data for dissolved oxygen concentration, and it was found that the error is around 1~3% [Lee 2018]. The full suite of equations [Lee 2018] derived from this basic model is reiterated below as:

$$K_L a = \frac{[1 - \exp(-K_L a_0 x(1-e)Z_d)]}{x(1-e)Z_d} \tag{5.3}$$

$$y = \frac{C}{nHP} + \left(\frac{Y_0 P_d}{P} - \frac{C}{nHP} \right) \exp(-xK_L a_0 \cdot mz) \tag{5.4}$$

$$C_\infty^* = nHY_0 \frac{P_a - P_d \exp(-mx \cdot K_L a_0 \cdot Z_d)}{1 - \exp(-mx \cdot K_L a_0 \cdot Z_d)} \tag{5.5}$$

$$K_L a = \frac{1 - \exp(-mx \cdot K_L a_0 \cdot Z_d)}{nmx \cdot Z_d} + \frac{(n-1)K_L a_0}{n} \tag{5.6}$$

$$Z_e = \frac{1}{mxK_L a_0} \left\{ \ln\left(P_e \frac{mxK_L a_0}{nr_w} \right) + \ln\left(\frac{nHY_0 P_d}{C_\infty^*} - 1 \right) \right\} \tag{5.7}$$

Finally, temperature has an effect on the value of $K_L a$, and the solution for temperature correction to standard conditions is given by Lee [2017] and in Chapter 2 Eq. 2.35 as:

$$(K_L a)_{20} = K_L a \frac{(E\rho\sigma)_{20}}{(E\rho\sigma)_T} \left(\frac{T_{20}}{T} \right)^5 \frac{P_{a_T}}{P_{a20}} \tag{5.8}$$

where E, ρ, σ are properties of the water under aeration. The model has its most precise application when used for shallow tanks or bench-scale experiments, at atmospheric pressures. All symbols are given in the Notation and also in previous paper [Lee 2018] or Chapter 2.

5.2   METHODOLOGY FOR DEPTH CORRECTION

The following steps lay out the procedure for calculating the specific baseline, and the flowchart for this procedure is as shown in Fig. 5.1 below:

(i) First, clean water tests (CWT) are to be done for 2 ~ 3 temperatures, preferably one below 20°C, one at 20°C and one above. CWT is also required for 2 different gas flow rates, so that altogether a minimum of 4 tests are recommended for a tank of adequate size, and the tank water depth is suggested to be fixed at 3 m (10 feet) or 5 m (15 feet) up to 7.6 m (25 ft). Tests will be repeated several times (minimum 3 as per ASCE standard) to have a constant $K_L a$ for each temperature, and the test is to be repeated for different applied gas flow rates (minimum number of gas flow rates is 2, since the point of origin constitutes a valid data point) so that the $K_L a$ vs. Q_a relationship can be estimated.

(ii) All diffused aeration systems will experience gas-side depletion as the water depth increases. This change in gas-side depletion is dealt with by the Lee-Baillod model [Lee 2018], allowing calculation of the baseline $(K_L a_0)$ by solving 5 simultaneous equations (Eq 5.3 to Eq. 5.7) (using the Microsoft Excel Solver or similar), where $K_L a_0$ is a variable to be determined, with the measured $K_L a$ and C_∞^* as the independent variables.

(iii) Once the baseline $K_L a_0$ for every test is established, a specific baseline can be determined using the $K_L a_0$ versus Q_a relationship. For the tests not done at 20°C and 1 atmosphere (atm) pressure, a temperature correction model to convert the specific baseline to standard conditions is required. The temperature correction model of choice is the fifth power model (Lee 2017) as advocated by the author. This *standardized* value can be used to find the transfer coefficient at another tank depth. By solving the simultaneous equations again, but using the established specific baseline parameter $(K_L a_0)_{20}$ as the independent variable, together with the actual environmental conditions surrounding the scaled-up tank, the $K_L a$ for the simulated tank can be found. Hence, the same set of equations are used twice, to calculate both the $K_L a_0$ and the $K_L a$.

(iv) All the measured apparent $K_L a$ values can be used to formulate the relationship between $K_L a$ and Q_a, but the resultant slopes may have some differences. This should be compared with the plot of $(K_L a_0)_T$ vs. Q_{aT} and also $(K_L a_0)_{20}$ vs. Q_{a20}. The latter curve should give the best correlation. Likewise, all $(K_L a_0)_T$ values are to be plotted against their respective handbook solubilities $(C_S)_T$.

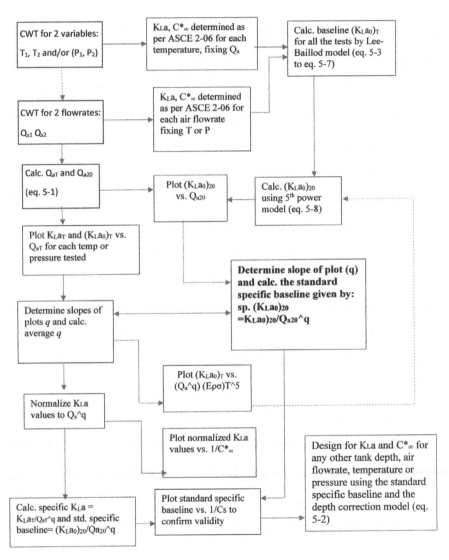

FIGURE 5.1 Flowchart for calculating the standard specific baseline $(K_L a_0)_{20}$

The specific baseline $(K_La_0)_{20}/Q_{a20}{}^q$ is expected to be constant for all the tanks tested. From the standardized baseline $(K_La_0)_{20}$ at 20°C, a family of rating curves for the standard mass transfer coefficient $(K_La)_{20}$ can thus be constructed for various gas flow rates applied to various tank depths using Eq. (5.2). Note that the standard specific baseline can also be expressed in terms of the superficial gas velocity U_g by simply dividing the average gas flowrate Q_a by the cross-section area S of the aeration column. It is difficult to describe a required geometry or placement for testing conducted in tanks other than the full-scale field facility. According to the ASCE Standard, appropriate configurations for shop tests should simulate the field conditions as closely as possible. For example, width-to-depth or length-to-width ratios should be similar. Potential interference resulting from wall effects and any extraneous piping or other materials in the tank should be minimized.

The density of the aerator placement, air flow per unit volume (or area), and power input per unit volume are examples of parameters that can be used to assist in making comparative evaluations. Notwithstanding these difficulties, the work here is to prove that, for the same configurations of aerator placement and tank dimensions, the model is able to predict oxygen transfer efficiency for a range of tank water depths and/or a range of other testing conditions, using a universal standard specific baseline mass transfer coefficient $(K_La_0)_{20}$. Conventional modeling uses the gas flow rate at standard conditions (Q_S) [Hwang and Stenstrom 1985], but since Q_S is in fact a *mass* flow rate rather than volumetric, K_La may not correlate well with Q_S. For any fixed gas supply rate of Q_S, K_La can be highly variable, such as in Case Study No. 1 described below, where the major variable on which K_La is dependent is the overhead pressure. There is no relationship between K_La and Q_S, but a strong correlation can be found between the *average* volumetric gas flow rate, Q_a, as calculated by Eq. 5.1, and K_La. The hypothesis advocated is that the baseline K_La_0 can be correlated with the volumetric gas flow rate Q_a as well. The fact that the mass transfer coefficient is an exponential function of the baseline means that the relationship between the former and gas flow rate would be different from that between the baseline and gas flow rate. The hypothesis in this manuscript is that the latter would constitute a better correlation, as can be seen in the case studies below, and is a power function.

5.3   MATERIALS AND METHODS

5.3.1   Case Study 1 - Super-oxygenation Tests

This research [Barber 2014] aimed to determine oxygen transfer rates, mass transfer coefficients, and saturation concentrations in clean water at different overhead pressures for a sealed aeration column, using air as well as high purity oxygen as the sources of oxygen. The experimental groups were designed to increase headspace pressures incrementally by 0.5 atm intervals, up to a pressure of 3 atmospheres, as shown in Table 5.1 below (note that LPM stands for liters per minute). The aeration apparatus was a clear acrylic tubular column totaling 5.64 m in height and 238 mm in diameter of a circular cross-section. The horizontal area is therefore given by

$S = 0.0445$ m^2. The column was fitted with a lid and O-ring to create an air-tight seal. After filling up with the test fluid but leaving a 0.3 m gap at the headspace, the column was pressurized to a desired pressure at the headspace. Sealed through the lid were three dissolved oxygen measuring probes—one near the surface of the water, one at mid-depth, and the third at near the bottom.

TABLE 5.1 Summary of the experimental design

Headspace Pressure (atm)	Gas Type	Gas Flowrate (LPM)	No. of Experiments	No. of Replicates	Total
1	Air/Oxygen	4	2	4	8
1.5	Air/Oxygen	4	2	4	8
2	Air/Oxygen	4	2	4	8
2.5	Air/Oxygen	4	2	4	8
3	Air/Oxygen	4	2	4	8
Total					40

A temperature probe was fitted at the mid-depth level to measure the water temperature during each test. For the air/oxygen supply, two 140 micron-air diffusers placed at the bottom and connected by air-hose flexible pipe 6.4 mm dia. that runs to the top of the column are connected via a drilled hole in the lid to the aeration feed-gas supply system.

5.3.1.1   Test Result for Air Aeration Tests

The aeration tests were carried out in accordance with The American Society of Civil Engineers Standard 2-06 [ASCE 2007] that requires a minimum of 3 replicates for non-steady state reaeration tests. However, in this experiment 4 replicates were provided for each probe, and 12 replicates for each test pressure. The reported values for $(K_La)_{20}$ and $C^*_{\infty 20}$ are given in column 3 and 5 as shown in Table 5.2 (note that N is the number of replicate tests). Experiments were conducted as close to the standard temperature of 20°C as possible, and the Arrhenius equation was used to the mass transfer coefficient K_La to standard conditions whenever an experiment was not conducted at that temperature. The actual temperatures were not reported, but given the range of temperatures reported as 10°C ~ 20°C for the air tests, and 15°C ~ 21°C for the HPO (high purity oxygen) tests, the Arrhenius model should be quite accurate; in any case, the effect of pressures must be greater than the effect from temperature discrepancies. In Table 5.3, Y_e is the oxygen mole fraction at the equilibrium position de, which is defined in ASCE 2-06 Standard as the effective depth; e is the ratio of the effective depth to the diffuser depth; and P_e is the equilibrium pressure at the effective depth. Note that Y_e can exceed 0.2095 at higher pressures, and the baseline becomes closer to the measured K_La simply because the overhead pressure dominates the effect on K_La rather that the water depth. The increase in P_e comes mainly from the overhead pressure although the hydrostatic pressure also contributes. In the above suite of equations (Eq. 5.3 to Eq. 5.7), the initial oxygen mole fraction Y_0 is 0.2095 for air, and 0.80 for the HPO tests.

TABLE 5.2 Data of the test results. (Number in parenthesis is +/- standard errors of the mean) (Symbols: bot = bottom; mid = mid-depth; surf = surface)

Z_d 5.33 m	Probe No.	$(K_La)_{20}$ (1/hr)	$(K_La)_{20}$ (1/min)	$C^*_{\infty 20}$ (mg/L)	SOTR (kg/hr)	SAE (kg/Kwh)	SOTE (%)	N
1 atm	1 (bot)	6.94 (0.2)	0.1157	11.62 (0.07)	19.48	1079	26.81	4
	2 (mid)	6.55 (0.1)	0.1092	11.38 (0.08)	18.00	996	24.76	4
	3 (surf)	6.49 (0.16)	0.1082	11.13 (0.08)	17.45	966	24.01	4
Average		*6.66 (0.1)*	*0.1110*	*11.38 (0.07)*	*18.31*	*1014*	*25.19*	*12*
1.5 atm	1 (bot)	4.84 (0.41)	0.0807	19.2 (1.31)	22.11	1078	30.83	4
	2 (mid)	5.42 (0.29)	0.0903	16.82 (0.36)	22.01	1073	30.68	4
	3 (surf)	4.92 (0.43)	0.0820	16.17 (0.63)	19.36	943	26.95	4
Average		*5.06 (0.21)*	*0.0843*	*17.39 (0.60)*	*21.16*	*1031*	*29.49*	*12*
2 atm	1 (bot)	4.39 (0.30)	0.0732	24.06 (1.97)	25.25	1122	35.61	4
	2 (mid)	4.81 (0.34)	0.0802	21.2 (0.32)	24.64	1095	34.75	4
	3 (surf)	3.93 (0.49)	0.0655	19.49 (1.09)	18.24	811	25.74	4
Average		*4.38 (0.23)*	*0.0730*	*21.58 (0.89)*	*22.71*	*1009*	*32.03*	*12*
2.5 atm	1 (bot)	4.35 (0.08)	0.0725	30.36 (0.72)	31.57	1288	44.37	4
	2 (mid)	3.38 (0.50)	0.0563	31.43 (2.42)	24.58	1003	34.55	4
	3 (surf)	2.83 (0.15)	0.0472	28.71 (1.44)	19.25	786	27.06	4
Average		*3.52 (0.25)*	*0.0587*	*30.17 (0.94)*	*25.13*	*1026*	*35.33*	*12*
3 atm	1 (bot)	3.59 (0.19)	0.0598	34.56 (1.13)	29.85	1156	42.67	4
	2 (mid)	3.17 (0.71)	0.0528	29.98 (0.65)	22.99	888	32.76	4
	3 (surf)	3.46 (0.42)	0.0577	36.98 (3.44)	31.64	1232	45.45	4
Average		*3.41 (0.26)*	*0.0568*	*33.84 (1.41)*	*28.16*	*1092*	*40.29*	*12*

The baseline values were calculated by solving the simultaneous equations (Eq. 5.3 to Eq. 5.7). The calculated values of $(K_La_0)_{20}$ are given in column 4 of Table 5.3. At higher pressures, it is not surprising that nitrogen gas is no longer inert (i.e. it participates in gas exchange) even though the water may already contain much nitrogen.

The Data Analysis Result for the air aeration is given in Table 5.3 below:

TABLE 5.3 Results of calculations of $K_L a_0$ (note that $S = 0.0445$ m^2)

Test Pressure	Q_S (L/min)	Q_a (m^3/min)	$K_L a_0$ (1/min)	Handbook Solubility C_S (mg/L)	Y_e	$e = d_e/Z_d$	P_e (N/m^2)
1 atm	4	0.0033	0.1254	9.09	0.2096	0.53	
			0.1161	9.09	0.2073	0.50	
			0.1155	9.09	0.2047	0.48	
Average			*0.1190*	*9.09*	*0.2072*	*0.50*	*125281*
1.5 atm	4	0.0023	0.0898	13.64	0.2095	0.56	
			0.0911	13.64	0.2136	0.58	
			0.0890	13.64	0.2093	0.51	
Average			*0.0900*	*13.64*	*0.2108*	*0.55*	*181875*
2 atm	4	0.0018	0.0724	18.18	0.2259	0.82	
			0.0808	18.18	0.2114	0.55	
			0.0750	18.18	0.2025	0.37	
Average			*0.0761*	*18.18*	*0.2133*	*0.58*	*230478*
2.5 atm	4	0.0015	0.0657	22.73	0.2297	0.97	
			0.0630	22.73	0.2342	1.06	
			0.0611	22.73	0.2230	0.82	
Average			*0.0633*	*22.73*	*0.2290*	*0.95*	*297622*
3 atm	4	0.0012	0.0548	27.27	0.2251	0.93	
			0.0546	27.27	0.2090	0.49	
			0.0597	27.27	0.2362	1.07	
Average			*0.0564*	*27.27*	*0.2234*	*0.83*	*344773*

This dissolution of nitrogen has an effect on Y_e, as the calibration factors n and m will change in response to nitrogen depletion in the gas stream. In fact, the Constant Bubble Volume Model (CBVM) was derived based on a particular scenario that the nitrogen gas in the bubble is being depleted as it rises to the surface, in such a way that the expansion of the bubble volume due to decreasing pressure is balanced by the reduction in volume due to loss of the nitrogen, so that the volume remains constant, and the calibration factors (n, m) are unity [Lee 2018]. Figure 5.2 shows the resulting $K_L a_0$ values (assumed to be at 20°C) plotted against the superficial gas velocity (also assumed to be at 20°C), from which the standard specific baseline $(K_L a_0)_{20}/U_{g20}{}^q$ is determined. The value obtained from the slope is 0.840 min^{-1}/(m/min)$^{0.75}$ for the air aeration. This standard specific baseline value so determined can then be used to simulate aeration tests for other conditions as described by Lee (2018). The oxygen-in-air solubility value (C_S) in Table 5.3, as stated in column 5, is obtained from Table 2.5 from Chapter 2. Previously, Lee (2017) has established that there is a definite inverse linear relationship between $K_L a$ and C^*_∞, provided that the test is done on a *very shallow* tank. The use of the baseline is tantamount to testing on a very shallow tank, and so, the same relationship would be expected to hold between $K_L a_0$ and solubility C_S. Since the relationship between the baseline and the gas flow rate is a power curve (Fig. 5.2), the relationship between the baseline and the inverse of solubility (or any other related function, if it exists, such as pressure) would also be a power curve, as shown by Fig. 5.3, because the baseline depends on the gas flow rate and varies with it as a power function. (For the sake of comparison, this plot also includes plotting the mass transfer coefficients $K_L a$ against the same insolubility values and it can be seen the relationship is not as good as the baseline plot relationship). The gas flow rate is, in turn, dependent on the hydrostatic pressure at the point of flow measurement P_P (see Eq. 5.1), which is dependent on tank height. Therefore, the *specific* baseline (i.e. $K_L a_0$ normalized to gas flow rate and pressure) versus the inverse of solubility, C_S, would form a straight line passing through the origin (as shown in Fig. 5.3a), as both parameters are then *independent* of tank height, and so, the inverse law between these two parameters ($K_L a_0$ vs. C_S) would be obeyed (Lee 2017). Note that $K_L a_0$ is defined at 1 atm pressure only. (For example, from Table 5.3, $K_L a_0(n) = .0611 * (.0033/.0015)^{0.75} * (1/2.5) = .044$ min^{-1} corresponding to $C_S = 22.73$ mg/L or $1/C_S = .044$ L/mg).

Figure 5.3b shows what happens when the saturation concentration is plotted against the overhead pressure. Barber (2014) also made the same plot as shown in Figure 3.2 in his report. At the same time, he calculated the saturation concentration using Henry's Law constant and the partial pressure at mid-depth, and found that the relationship is not linear. However, because of gas-side oxygen depletion, the equilibrium pressure may not be at mid-depth. When $C^*_{\infty 20}$ is plotted against P_e, the correlation improves as shown in the second curve from top. Furthermore, Henry's Law only applies to gas solubility, *not* to saturation concentration. The bulk liquid is actually "super-saturated" in aeration, and this saturation value may not relate to pressure in accordance with Henry's Law. When the oxygen solubility is plotted against the overhead pressure, then Henry's Law would apply and this is shown as a perfect straight in the bottom curve in Fig. 5.3b. Similar to Figure 3.1 in Barber's report, the standard mass transfer coefficient is plotted against the pressure and obtains the similar finding as shown in Fig. 5.3c. However, when the baseline plot is superimposed on this graph, it can be seen that the correlation between the mass transfer coefficient (baseline) and the headspace pressure becomes much better. When it is plotted against the inverse of the pressures, it can be seen that a power curve is obtained as shown in Fig. 5.3d. This confirms the hypothesis of Eq. 2.1 in Chapter 2 where it was postulated that $K_L a$ is inversely proportional to P_S but only for shallow tanks, which is equivalent to plotting $K_L a_0$ against the inverse of pressure P_S. When the *baseline* is normalized to *the gas flow rate to power q*, a straight line linear relationship would be obtained for such correlation (plot not shown but would be similar to Fig. 5.3a for solubility). Similar findings on the inverse relationship between the mass transfer coefficient ($K_L a$) and saturation concentration (C^*_∞), as well as the inverse relationship between the baseline mass transfer coefficient $(K_L a_0)$ and pressure (P_S), can be found for high purity oxygen aeration, as shown in Figs. 5.5, 5.5a, and 5.5b.

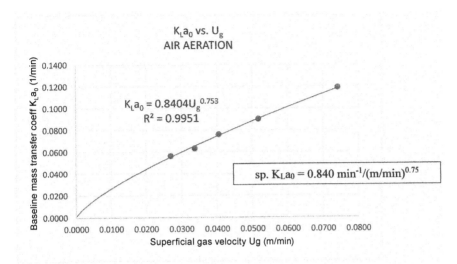

FIGURE 5.2 Standard baseline $(K_L a_0)_{20}$ versus average standard gas velocity U_g for various test pressures

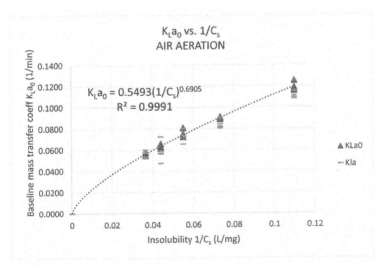

FIGURE 5.3 Standard baseline $(K_L a_0)_{20}$ versus inverse of solubility $1/C_S$ for various pressures

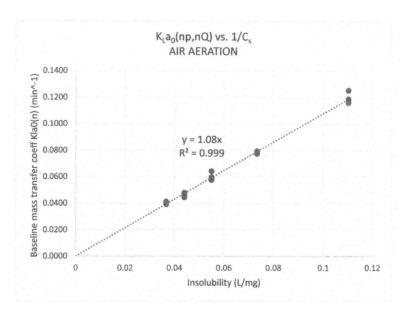

FIGURE 5.3a Standard baseline $(K_L a_0)_{20}$ normalized to P and Q versus inverse of solubility $1/C_S$ for various pressures

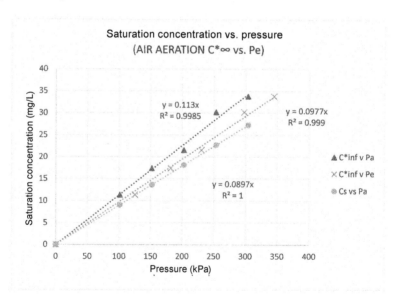

FIGURE 5.3b Saturation concentration for air for various pressures (note that Henry's law is given by the third curve from the top)

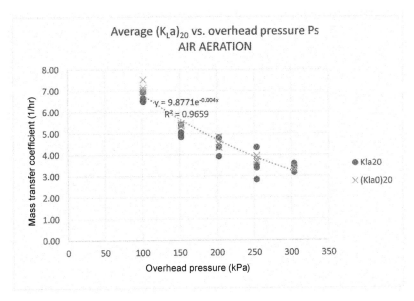

FIGURE 5.3c Standard baseline $(K_La_0)_{20}$ compared with mass transfer coefficient $(K_La)_{20}$ for various pressures

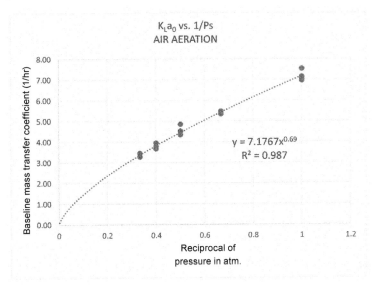

FIGURE 5.3d Standard baseline $(K_La_0)_{20}$ versus inverse of pressures $1/P_S$

5.3.1.2 High Purity Oxygen (HPO) Aeration Test Result

The reported HPO test values for $(K_La)_{20}$ and $C^*_{\infty 20}$ are given in column 3 and 5 as shown in Table 5.4 below. Table 5.5 shows the analysis result of the HPO tests. Figure 5.4 shows the resulting K_La_0 values (assumed to be at 20°C) plotted against the superficial gas velocity (also assumed to be at 20°C), from which the standard specific baseline $K_La_{020}/U_{g20}{}^q$ is determined. The value obtained from the slope is

1.511 min^{-1}/(m/min)$^{0.92}$ which is the specific baseline for the HPO aeration. This value can then be used to simulate aeration tests for other conditions as described by Lee (2018). As mentioned before, since the relation between the mass transfer coefficient and the gas flow rate is a power curve, and since the baseline K_La is inversely proportional to oxygen solubility C_S, the plot of the baseline versus the inverse of solubility would also give a power curve, as shown in Fig. 5.5. Pure oxygen solubility values in water is given in col. 5 in Table 5.5. Similar argument can be applied to the HPO tests as to the air aeration tests, giving, therefore, the plots as shown in Fig. 5.4 and Fig. 5.5, showing the excellent correlation between the baseline and the reciprocal of oxygen solubility in the latter plot. The relationship is a power curve, as is the relationship between the baseline and the gas flow rate (Fig. 5.4). As would be expected, the *specific* baseline (i.e. K_La_0 normalized to gas flow rate and pressure) versus the inverse of solubility, C_S, would form a straight line passing through the origin (plot not shown), as both parameters are then *independent* of tank height, and so, the inverse law between these two parameters would be obeyed (Lee 2017). For comparison, the measured apparent mass transfer coefficients are also plotted against the inverse of solubility in Fig. 5.5, and it can be seen that the correlation is not as good compared to the baseline plot, further confirming the model validity for the baseline.

TABLE 5.4 Data of the pure oxygen test results. (Number in parenthesis is +/− standard errors of the mean) (Note that $S = 0.0445$ m^2)

Z_d 5.33 m	Probe No.	$(K_La)_{20}$ (1/hr)	$(K_La)_{20}$ (1/min)	$C^*_{\infty20}$ (mg/L)	SOTR (kg/hr)	SAE (kg/Kwh)	SOTE (%)	N
1 atm	1 (bot)	7.85 (0.50)	0.1308	51.37 (0.32)	97.75	241.0	35.12	4
	2 (mid)	8.13 (0.28)	0.1355	49.63 (0.42)	97.89	241.3	35.17	4
	3 (surf)	8.27 (0.30)	0.1378	49.17 (0.33)	98.62	243.1	35.43	4
Average		*8.08 (0.20)*	*0.1347*	*50.06 (0.34)*	*98.09*	*241.8*	*35.24*	*12*
1.5 atm	1 (bot)	5.51 (0.29)	0.0918	75.30 (2.19)	99.67	223.5	38.44	4
	2 (mid)	5.44 (0.55)	0.0907	74.15 (1.33)	97.15	218.2	37.40	4
	3 (surf)	5.52 (0.43)	0.0920	71.07 (0.65)	94.68	212.6	36.45	4
Average		*5.49 (0.23)*	*0.0915*	*73.51 (0.96)*	*97.17*	*218.1*	*37.43*	*12*
2 atm	1 (bot)	4.51 (0.28)	0.0752	91.84 (1.56)	97.88	204.0	39.97	4
	2 (mid)	5.38 (0.21)	0.0897	87.14 (2.33)	110.80	231.0	45.38	4
	3 (surf)	5.26 (0.25)	0.0877	81.05 (2.54)	100.60	209.6	41.20	4
Average		*5.05 (0.17)*	*0.0842*	*86.68 (1.75)*	*103.10*	*214.9*	*42.18*	*12*
2.5 atm	1 (bot)	3.07 (0.22)	0.0512	106.0 (5.73)	77.76	152.9	32.68	4
	2 (mid)	4.70 (0.26)	0.0783	94.70 (5.71)	106.40	208.8	44.57	4
	3 (surf)	4.53 (0.69)	0.0755	91.73 (1.99)	100.20	197.7	42.15	4
Average		*4.10 (0.26)*	*0.0683*	*97.46 (2.17)*	*94.75*	*186.5*	*39.80*	*12*
3 atm	1 (bot)	4.07 (0.07)	0.0678	119.5 (3.10)	115.10	217.8	45.59	4
	2 (mid)	3.02 (0.54)	0.0503	123.7 (6.78)	85.22	161.6	35.49	4
	3 (surf)	3.68 (0.17)	0.0613	106.3 (2.15)	92.58	175.5	36.72	4
Average		*3.59 (0.18)*	*0.0598*	*116.5 (2.90)*	*97.63*	*185.0*	*39.27*	*12*

TABLE 5.5 Results of calculations of $K_L a_0$ for the HPO tests (note that $S = 0.0445$ m^2)

Test Pressure	Q_S (L/min)	Q_a (m^3/min)	$K_L a_0$ (1/min)	Handbook Solubility C_S (mg/L)	Y_e	$e = d_e/Z_d$	P_e (N/m^2)
1 atm	4	0.0033	0.1423	34.71	0.7773	0.99	
			0.1429	34.71	0.7509	0.99	
			0.1431	34.71	0.7440	0.99	
Average			*0.1428*	*34.71*	*0.7574*	*0.99*	*152979*
1.5 atm	4	0.0023	0.0987	52.09	0.8534	0.99	
			0.0989	52.09	0.8404	0.99	
			0.0995	52.09	0.8055	0.99	
Average			*0.0990*	*52.09*	*0.8331*	*0.99*	*203641*
2 atm	4	0.0018	0.0774	69.42	0.8320	0.99	
			0.0780	69.42	0.7895	0.99	
			0.0788	69.42	0.7343	0.99	
Average			*0.0781*	*69.42*	*0.7853*	*0.99*	*254304*
2.5 atm	4	0.0015	0.0693	86.80	0.7999	0.99	
			0.0658	86.80	0.7146	0.99	
			0.0661	86.80	0.6922	0.99	
Average			*0.0671*	*86.80*	*0.7356*	*0.99*	*304966*
3 atm	4	0.0012	0.0554	104.13	0.7726	0.98	
			0.0551	104.13	0.7998	0.98	
			0.0596	104.13	0.6873	0.98	
Average			*0.0567*	*104.13*	*0.7532*	*0.98*	*355107*

In Fig. 5.5a, the mass transfer coefficients are plotted against the inverse of C_∞^*, which gives a reasonably good fit, but are always inferior to the baseline plot given in Fig. 5.5. Similarly, when the baseline is plotted against the inverse of pressure, an

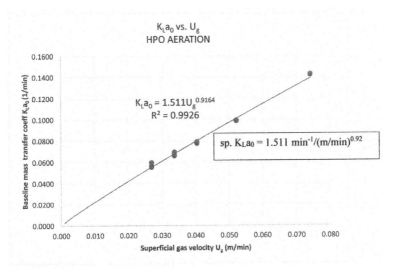

FIGURE 5.4 Standard baseline $(K_L a_0)_{20}$ versus average standard gas velocity U_g

excellent correlation is obtained, as shown in Fig. 5.5b. This curve is similar to Fig. 5.3d for the air aeration, thus validating the proposed model in Chapter 2, as stated by Eq. 2.35 for the pressure correction, and also by Eq. 5.8 above.

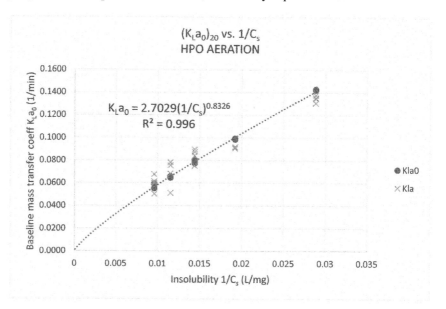

FIGURE 5.5 The inverse relationship between baseline K_La_0 and solubility $1/C_S$ for various pressures

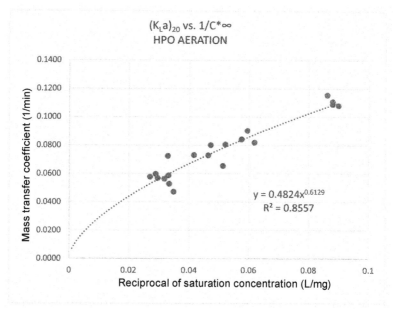

FIGURE 5.5a Standard mass transfer coefficient $(K_La)_{20}$ versus inverse of saturation concentration $1/C^*_\infty$ for various pressures

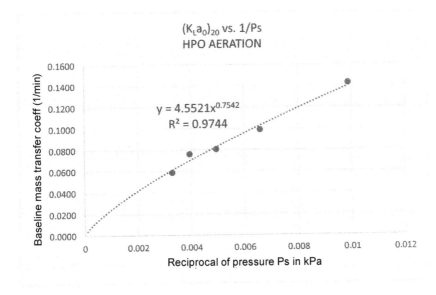

$(K_La_0)_{20}$ vs. 1/Ps
HPO AERATION

$y = 4.5521x^{0.7542}$
$R^2 = 0.9744$

Baseline mass transfer coeff (1/min)

Reciprocal of pressure Ps in kPa

FIGURE 5.5b Standard baseline $(K_La_0)_{20}$ versus inverse of pressure $1/P_S$

5.3.2 Case Study 2 - ADS (Air Diffuser Systems) Aeration Tests

Testing was performed by Stenstrom (2001) in a clear acrylic column with a 222 mm (8.75 in.) internal diameter by 4.88 m (16 ft) maximum depth. The depth of the column was varied by filling it to different heights with tap water to the appropriate depth. Small sections of aeration tubing with slits were placed at the bottom to create orifices for aeration. The reported test results for the standard mass transfer coefficient and the saturation concentration are given in Table 5.6, with the data analysis results and the calculated baseline values for the mass transfer coefficients. The test water came from Lake Bluff tap water, and since the tests were done in early March, is expected to be below 20°C. The tests were performed and the data analyzed in strict adherence to the ASCE standard, and, therefore, the Arrhenius temperature-correction model was used. The correction is deemed to be accurate because the model is known to work well for water temperatures below 20°C within the range of 10°C and 30°C stipulated by the standard.

Two types of ADS Aeration Tubing were tested, one has 6 slits and the other had 14 slits. Tests were done at three depths: 1.52 m (5 feet), 3.05 m (10 feet), and 4.57 m (15 ft). Three probes were deployed in the last two depths. Three tests were performed at the design flow rate for eachorifice configuration in order to provide a measure of the precision that is required by the Standard. Figure 5.6 shows the resulting K_La_0 values (assumed to be at 20°C) plotted against the superficial gas velocity (also assumed to be at 20°C), from which the standard specific baseline $K_La_{020}/U_{g20}{}^q$ is determined. The value obtained from the slope is 3.256 hr^{-1}/(m/hr)$^{0.71}$ or **0.994 min^{-1}/(m/min)$^{0.71}$**, which is the specific baseline for the air aeration. This value can then be used to simulate aeration tests for other conditions as described by Lee (2018).

TABLE 5.6 Results of calculations of $K_L a_0$ (note that $S = 0.0387$ m²)

Run No.	Water Dep (m)	AFR (scmh)	$C^*_{\infty 20}$ (mg/L)	Q_a (m3/h)	U_g (m/h)	$(K_L a_0)_{20}$ (1/h)	$(K_L a)_{20}$ (1/h)	y_e	sp. $(K_L a_0)_{20}$
4	4.57	0.0401	10.61	0.034	0.89	3.16	2.73	0.1979	3.44
1	4.57	0.0802	10.67	0.069	1.77	6.26	5.52	0.1988	4.17
2	4.57	0.0802	10.67	0.069	1.77	6.27	5.53	0.1988	4.18
3	4.57	0.0802	10.79	0.069	1.77	6.20	5.49	0.2002	4.13
5	4.57	0.1604	10.75	0.138	3.55	7.39	6.76	0.1998	3.01
12	3.05	0.0401	10.04	0.036	0.93	3.16	2.86	0.2012	3.34
9	3.05	0.0802	10.08	0.072	1.85	4.69	4.35	0.2022	3.03
10	3.05	0.0802	10.06	0.072	1.85	4.67	4.34	0.2020	3.02
11	3.05	0.0802	10.06	0.072	1.85	4.58	4.27	0.2020	2.96
13	3.05	0.1604	10.13	0.144	3.70	6.70	6.33	0.2029	2.64
23	1.52	0.0401	9.56	0.038	0.97	3.64	3.38	0.2096	3.71
20	1.52	0.0802	9.489	0.076	1.95	5.55	5.27	0.2094	3.46
21	1.52	0.0802	9.473	0.076	1.95	5.52	5.24	0.2092	3.44
22	1.52	0.0802	9.526	0.076	1.95	5.35	5.08	0.2100	3.33
24	1.52	0.1604	9.525	0.151	3.89	7.75	7.45	0.2104	2.95
6	4.57	0.0344	10.94	0.030	0.76	2.31	2.01	0.2019	2.81
7	4.57	0.0344	10.93	0.030	0.76	2.60	2.23	0.2018	3.15
8	4.57	0.0344	10.9	0.030	0.76	2.45	2.11	0.2014	2.97
14	3.05	0.0344	10.18	0.031	0.79	2.69	2.42	0.2032	3.17
15	3.05	0.0344	10.16	0.031	0.79	2.62	2.35	0.2030	3.08
16	3.05	0.0344	10.17	0.031	0.79	2.63	2.37	0.2031	3.10
17	1.52	0.0344	9.504	0.032	0.83	2.81	2.64	0.2091	3.20
18	1.52	0.0344	9.507	0.032	0.83	2.78	2.61	0.2092	3.16
19	1.52	0.0344	9.463	0.032	0.83	2.78	2.61	0.2086	3.16

Since the airflow rate for the 14-slit tubing was 0.0802 scmh (0.0472 scfm), and that for the 6-slit was 0.0344 scmh (0.0203 scfm), where scmh stands for standard cubic meters per hour, and scfm stands for standard cubic feet per minute, with the 14-slit additionally tested at 50% and 200% of the design flow rates to determine the impact of airflow rate on oxygen transfer efficiency, there may be some discrepancies in the baseline calculations, especially at the higher gas flow rate values. This probably explains why the data at the top end of the graph shows some anomalies.

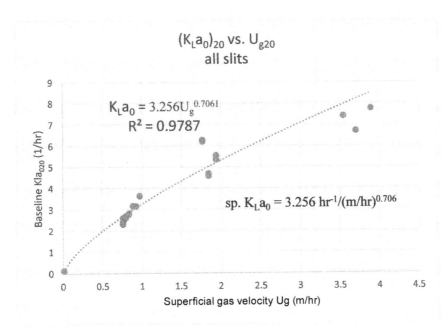

FIGURE 5.6 Standard baseline $(K_L a_0)_{20}$ versus average standard gas velocity U_g

5.3.3 Case Study 3 - FMC, Norton and Pentech Jet Aeration Shop Tests

The test facility used by Yunt et al. [1988a], Yunt and Hancuff [1988b] for all tests was an all steel rectangular aeration tank located at the Los Angeles County Sanitation Districts (LACSD) Joint Water Pollution Control Plant. The details of the test have already been described in Chapter 3 Section 3.3 for the FMC diffusers. This case study is included in this chapter for the purpose of completeness.

The Norton diffusers are fine bubble dome type, and consist of 126 ceramic dome diffusers mounted on PVC headers. The test results are given in the LACSD report Table 5: "Summary of Exponential Method Results: FMC Fine Bubble Tube Diffusers" and copied herewith as Table 5.7 (which is the same as Table 3.1) below. Similarly, Table 5.8 gives the test results for the Norton diffusers. Table 5.7 and 5.8 below are compiled based on data contained in the LACSD report for FMC and Norton Fine Bubble Diffusers. (A similar table for the Pentech Jet test data has also been compiled but not shown here). These tables show all the raw data as given in the LACSD report. The calculations for estimating the variables $K_L a_0$, n, m, d_e and y_e are not shown in this chapter, but the reader is referred to the example calculation of the baseline mass transfer coefficient $(K_L a_0)_T$ using the model equations in Chapter 3. The simulated result for the FMC diffusers in this example gives a value of $(K_L a)_{20} = 0.1874$ min^{-1} for a typical run test as compared to the test-reported value of $(K_L a)_{20} = 0.1853$ min^{-1} which gives an error difference of around 1% only comparing to the simulated value.

5.4   EXAMPLE CALCULATIONS

An example has been given in Chapter 3 Section 3.4 for the FMC diffusers. The test data are as shown in Table 5.7. Figure 5.7 (which is the same as Fig. 3.3) shows that the resulting $K_L a_0$ values are adjusted to the standard temperature by the temperature correction equation of the 5th power model (Lee 2017) and plotted against Q_{a20}. *These curves relating $K_L a_0$ with Q_a for each tank depth are all fitted together after normalizing $K_L a_0$ values to 20°C, as shown in the graph, to form one single curve. The exponent determined is 0.82.* The value obtained from the slope is **0.044 min⁻¹/ (m³/min)$^{0.82}$** or **0.861 min⁻¹/(m/min)$^{0.82}$** for all the gas rates normalized to give the best NLLS (Non-Linear Least Squares) fit, bearing in mind that the $K_L a_0$ is assumed to be related to the gas flowrate by a power curve with an exponent value [Rosso and Stenstrom 2006] [Zhou et al. 2013]. The slope of the curve is defined as the standard specific baseline. Therefore, the standard specific baseline (sp. $K_L a_0)_{20}$ is calculated by the ratio of $(K_L a_0)_{20}$ to Q_{a20} 82 or by the slope of the curve in Fig. 5.7. The graph output for the Pentech Jet is shown in Fig. 5.8a. Again, a standard specific baseline can be obtained for all the gas rates normalized to give the best NLLS (Non-Linear Least Squares) fit, since the $K_L a_0$ value is related to the gas flowrate by a power curve with an exponent q [Hwang and Stenstrom 1985] [Rosso and Stenstrom 2006] [Zhou et al. 2013].

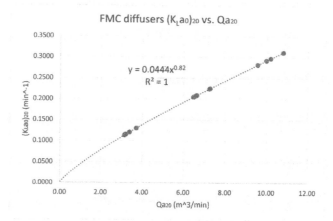

FMC diffusers $(K_L a_0)_{20}$ vs. Q_{a20}

$y = 0.0444x^{0.82}$
$R^2 = 1$

(KLa0)₂₀ (min^-1)

Qa20 (m^3/min)

FIGURE 5.7 Calculation of the standard specific baseline $(K_L a_0)_{20}/Q_{a20}^{0.82}$ for various test temperatures and water depths (FMC diffusers)

For the Pentech jets, the resulting baseline (with one outlier removed) is given as **0.0515 min⁻¹/(m³/min)$^{0.728}$** or **0.716 min⁻¹/(m/min)$^{0.71}$**. The graph for all the data including the outlier is given in Fig. 5.8b below. Therefore, it would seem that the methodology for calculating the baselines has the added benefit of spotting outliers in the testing data, as it seems obvious that the last data shown in the graph is an outlier. At the very least, it serves as a red flag that this last data point is questionable, enabling the researcher to re-visit this particular test and perhaps carry out the test once more to confirm its validity. For the Norton diffusers, the baseline is measured to be **0.072 min⁻¹/(m³/min)$^{0.80}$** or **1.305 min⁻¹/(m/min)$^{0.80}$** as shown in Fig. 5.8.

TABLE 5.7 Los Angeles county sanitation district report test data (1978) FMC diffusers

Date	#	Water Depth	Delivered Power Density	Air-flow Rate Q_s	Temperature	$(K_L a)_{20}$	$(K_L a)_{20}$	$C^*_{\infty 20}$	Standard Oxygen Transfer Efficiency
		m	(hp/ 1000 ft^3)	scmh	T (°C)*	1/hr	1/min	mg/L	(%)
Aug 29,78	1	3.05	2.02	700	25.2	17.46	0.2910	9.87	10.06
Aug 29,78	2	3.05	1.16	470	25.2	13.37	0.2228	9.99	11.68
Aug 29,78	3	3.05	0.54	241	25.2	7.63	0.1272	10.05	12.95
Aug 29,78	1	7.62	1.66	704	25.2	14.99	0.2498	11.23	23.93
Aug 29,78	2	7.62	1.07	478	25.2	11.12	0.1853	11.26	24.40
Aug 29,78	3	7.62	0.51	236	25.2	6.39	0.1065	11.54	31.71
Sep 29,78	1	3.05	1.19	472	25.2	13.39	0.2232	9.98	11.61
Feb 08,79	1	4.57	1.81	694	16.2	16.61	0.2768	10.50	15.34
Feb 08,79	2	4.57	1.05	449	16.2	11.90	0.1983	10.54	17.07
Feb 08,79	3	4.57	0.51	231	16.2	6.88	0.1147	10.63	19.87
Feb 08,79	1	6.10	1.74	709	16.2	16.73	0.2788	10.80	20.69
Feb 08,79	2	6.10	1.08	471	16.2	11.62	0.1937	11.05	22.17
Feb 08,79	3	6.10	0.49	224	16.2	6.10	0.1017	11.19	25.04

*Note: Water temperature was deduced from the report statement: "The temperature range used in the study was 16.2 to 25.2°C." Reported main data are given in bold; $(K_L a)_{20}$ given in this table is based on the Arrhenius model using

$$\theta = 1.024 \qquad \text{[ASCE 2007]}$$

The 5th power temperature correction model [Lee 2017] to convert $K_L a_0$ estimated to $(K_L a_0)_{20}$ and subsequently to $(K_L a)_{20}$ is given by:

$$(K_L a)_{20} = K_L a \frac{(E\rho\sigma)_{20}}{(E\rho\sigma)_T}\left(\frac{T_{20}}{T}\right)^5$$

Note that there were some discrepancies in the reported data for the 7 m tank, in that the data for the SOTE% were calculated by an equation in the report and they did not match up for two points in the report. These data were discarded and the calculated values using the LACSD Report's equations were used, but these data are still suspect. The greater the number of tests are done, the better would be the estimation of the unknown parameters.

TABLE 5.8 Los Angeles county sanitation district report test data (1978) Norton diffusers

Date	#	Water Depth	Delivered Power Density	Air-flow Rate Q_s	Temperature	$(K_La)_{20}$	$(K_La)_{20}$	$C^*_{\infty 20}$	Standard Oxygen Transfer Efficiency
		m	(hp/1000 ft^3)	scmh	$T(°C)^*$	1/hr	1/min	mg/L	(%)
Mar 24,78	1	7.62	0.28	125	20	5.34	0.0890	11.42	49.48
Apr 21,78	1	3.05	0.57	214	20	11.31	0.1885	9.81	21.30
Apr 24,78	1	3.05	0.32	126	20	7.17	0.1195	9.88	23.20
Apr 25,78	1	4.57	0.31	127	20	6.41	0.1068	10.24	32.03
Apr 26,78	1	4.57	0.54	214	20	9.87	0.1645	10.45	29.71
Apr 27,78	1	4.57	1.24	430	20	17.66	0.2943	10.6	26.61
May 04,78	1	6.10	0.51	216	20	9.47	0.1578	11.12	39.81
May 05,78	1	6.10	1.15	435	20	16.39	0.2732	11.02	33.80
May 08,78	1	7.62	1.16	463	20	14.61	0.2435	11.67	37.16
May 09,78	1	7.62	0.50	217	20	8.54	0.1423	11.65	46.69
May 10,78	1	6.10	0.30	130	20	6.07	0.1012	11.33	43.55
May 15,78	1	3.05	1.37	422	20	19.3	0.3217	10.17	19.14
May 16,78	1	6.10	0.30	127	20	5.82	0.0970	11.44	42.85

*Note: Water temperature was assumed to be 20°C, actual temperatures not reported. Reported main data are given in bold; $(K_La)_{20}$ given in this table is based on the Arrhenius model using

$$\theta = 1.024 \qquad\qquad [ASCE\ 2007]$$

Since the tests were carried out from March to May, the water temperature is likely to be 20°C or less, so that the Arrhenius model is likely to be quite accurate. The greater the number of tests are done, the better would be the estimation of the unknown parameters. In the absence of more data, all the tests are assumed to have been carried out under standard conditions.

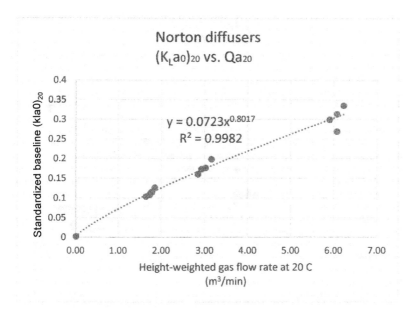

FIGURE 5.8 Calculation of the standard specific baseline $(K_L a_0)_{20}/Q_{a20}^{0.80}$ for various water depths (Norton diffusers)

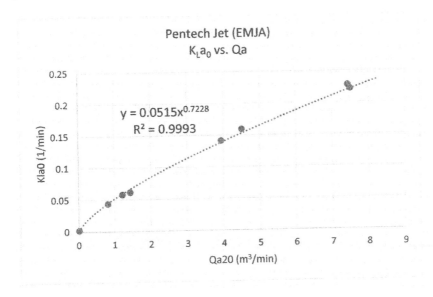

FIGURE 5.8a Calculation of the standard specific baseline $(K_L a_0)_{20}/Q_{a20}^{0.72}$ for various test temperatures and water depths (Pentech Jets) w/outlier removed

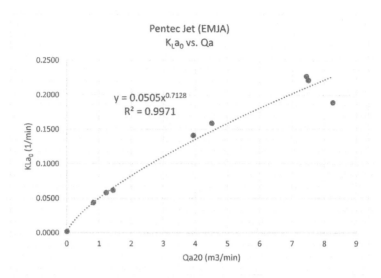

FIGURE 5.8b Calculation of the standard specific baseline $(K_L a_0)_{20}/Q_{a20}^{0.72}$ for various test temperatures and water depths (Pentech Jets) w/all data

5.5 DISCUSSION

5.5.1 Rating Curves for Aeration Equipment

The good prediction of the baseline mass transfer coefficient is a breakthrough since the correct prediction of the volumetric mass transfer coefficient $(K_L a)$ using the baseline is a crucial step in the design, operation and scale up of bioreactors including wastewater treatment plant aeration tanks, and the equation developed allows doing so without resorting to multiple full-scale testing for each individual tank under the same testing condition for different tank heights and temperatures.

As mentioned in the Methodology section, a family of rating curves for $(K_L a)_{20}$ with respect to depth can thus be constructed for various gas flow rates applied, such as the one shown in Fig. 5.9 and 5.10 below.

In Fig. 5.9, the rating curves were constructed based on the three average gas flow rates, which vary slightly for each tank depth, but the sp. $K_L a_0$ has been normalized to a constant gas flow rate for each curve. (Compare this graph with Fig. 3.8 which has used the test gas flow as is without any normalization, showing that the shape of the curves is sensitive to gas flow rate).

As mentioned in Chapter 3, although the rating curves show that the $(K_L a)_{20}$ values are always less than the baseline $(K_L a_0)_{20}$, it is generally accepted that the deeper the tank, the higher the oxygen transfer efficiency, all things else being equal [Houck and Boon 1980] [Yunt et al. 1988a], [Yunt and Hancuff 1988b]. This is simply because the dissolved oxygen saturation concentration increases with depth, which offsets the loss in the transfer coefficient in a deep tank. The net result is therefore still an increase in the overall aeration efficiency.

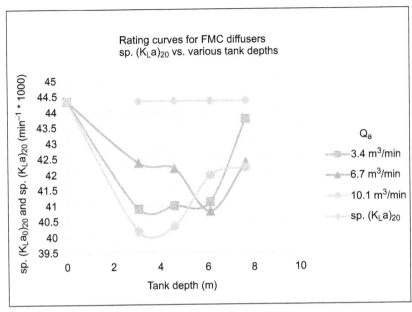

FIGURE 5.9 Rating curves for the standard specific transfer coefficients (K_La_0 and $K_La)_{20}$ for various tank depths and air flow rates (FMC diffusers)

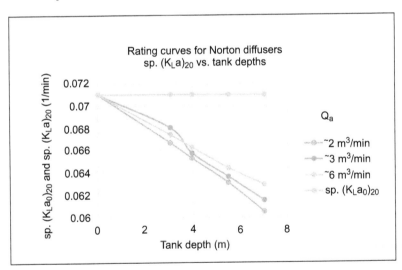

FIGURE 5.10 Rating curves for the standard specific transfer coefficients (K_La_0 and $K_La)_{20}$ for various tank depths and air flow rates (Norton diffusers)

Other clean water studies showed a nearly linear correlation between oxygen transfer efficiency and depth up to at least 6.1 m (20 ft) [Houck and Boon 1980]. The rating curves show that, in general, K_La decreases with depth for a fixed average volumetric gas flowrate. For the gas flowrates in Fig. 5.9, for example, the profile is almost linear up to 4 m, which confirms Downing and Boon's finding [Boon 1979]. For Norton diffusers, the curves are linear up to 7 m as shown in Fig. 5.10.

DeMoyer et al. (2002) and Schierholz et al. (2006) have conducted experiments to show the effect of free surface transfer on diffused aeration systems, and it was shown that high surface-transfer coefficients exist above the bubble plumes, especially when the air discharge rate (Q_a) is large. In the establishment of the baselines, care must be taken in selecting the test gas flow rates and tank geometry such that other such effects would not render the simulation model invalid. The simulation model has ignored any free surface effects [Lee 2018]. When coupled with large surface cross-sectional area and/or shallow depth, the oxygen transfer mechanism becomes more akin to surface aeration where air entrainment from the atmosphere becomes important. In order to make the model valid, the alternative to a judicious choice of tank geometry and/or gas discharge is perhaps another mathematical model that could separate the effect of surface aeration from the actual aeration under testing in the estimation of the baseline coefficient [DeMoyer et al. 2002] [Schierholz et al. 2006]. This topic would be the subject of another paper and is briefly discussed in Chapter 6 Section 6.5.4.

5.5.2   Justification of the 5th Power Model Over the ASCE Method for Temperature Correction

The advantage of the 5th power model for temperature correction is that it is a base model, around which other effects can be built on, such as the stirrer speed or rotation speed of impeller, gas flow rates, geometry, dissolved solids, liquid characteristics, and so forth, which can be accounted for, provided these effects' relationships with temperature are individually known. Since the effect of tank height has been accounted for in the study, it was thought that this model would give a more reliable estimate of $(K_L a_0)_{20}$ than the Arrhenius equation. By comparison, the Arrhenius relation, derived for the temperature dependence of the equilibrium constants of ideal gas mixtures and shown to fit data for the temperature relationship of many reaction rate constants, is used empirically for gas mass transfer. However, gas transfer is a diffusion process, not a chemical reaction. K_L is not a reaction rate coefficient and so the Arrhenius equation is not theoretically based. Therefore, the Arrhenius equation used in this context is purely empirical. Daniil and Gulliver (1988) have made a comparison between various temperature correction models, and concluded that the one using properties of water and derived using dimensional analysis incorporating the Schmidt number and others, has the best similarity with the θ model of 1.0241. Although they recommended replacing the Arrhenius equation with this dimensional equation, it was also recognized that their favored equation is not universal either, since the equation has not accounted for the turbulence effect and the effect of K_L itself on the value of θ.

Since the test temperature range falls within the ASCE prescribed temperature range (10°C ~ 30°C), the Arrhenius equation using $\theta = 1.024$ is also approximately valid (see Fig. 5.11 for comparison of various temperature correction models for the FMC diffusers). However, as the purpose of this manuscript is to advance a depth correction model, so that one test carried out at a certain tank depth can be translated to test results for other tank depths, the use of a 5th power model appears to give the best regression analysis yet to yield the standard baseline $(K_La_0)_{20}$.

The θ parameter attempts to lump all the effects together, and therefore it does not allow modifications to include other effects, except doing more experiments to suit each case, and altering the θ value altogether based on these experiments (Lee 2017). The θ parameter is an 'all-in' function of many effects, sometimes including temperature itself as explained in the paper "Temperature Effects in Treatment Wetlands" [Kadlec and Reddy 2001]. As shown in Fig. 5.11 for the FMC diffusers, the discrepancies between these models in terms of standardizing $(K_La_0)_T$ to $(K_La_0)_{20}$ are very small. Table 2.3 in Chapter 2 showed Vogelaar et al.'s data [Vogelaar et al. 2000] and the relationship between the $(K_La)_T$ and the inverse of $C^*_{\infty T}$ is plotted in Fig. 7.1 in Chapter 7, clearly illustrating the linear relationship between these two parameters for different temperatures. It should be noted in passing that Fig. 3.4 in Chapter 3 is irrelevant insofar as the main theme of the manuscript is concerned; its purpose is to only show that the relationship of the measured K_La (normalized to gas flow rate) with saturation concentration, although such relationship exists, is not as good as the same relationship using the baseline $K_La(K_La_0)$ instead, as given in Fig. 3.5, which has a higher correlation coefficient $R^2 = 0.9924$ compared to $R^2 = 0.9859$. (Note that when the baseline K_La_0 is used, the corresponding saturation concentration is reduced to surface saturation).

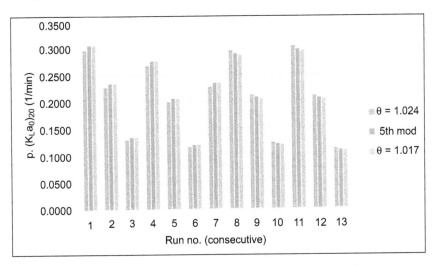

FIGURE 5.11 Comparison of the standard baseline $(K_La_0)_{20}$ using various temperature models

The farther the tank departs from a shallow tank, the more the deviation is for the mutual linear correlation between the two parameters (K_La vs. C^*_∞), according to the model given by Eq. (3.6) for the depth correction, shown in Fig. 3.10, which is an exponential function, contrary to the solubility (C_S) or the saturation concentration (C^*_∞) variation model with depth, which is linear in accordance with Henry's Law.

5.6 CONCLUSION

Oxygen is only slightly soluble in water. Therefore, the mass transfer coefficient K_La is extremely sensitive to the gas depletion rate in a bubble, which in turn is highly sensitive to changes in the factors affecting its depletion. By citing several case studies, this paper has illustrated that, for each specific submerged bubble aeration equipment, the standard baseline $(K_La_0)_{20}$ at the standard temperature of 20°C and standard pressure of 1 atm is a constant for any specific feed-gas composition, regardless of tank depth, test water temperature and overhead test pressure, when it is normalized to a fixed gas flowrate Q_a. This baseline value can be expressed as a specific standard baseline when the relationship between $(K_La_0)_{20}$ and the average volumetric gas flow rate Q_{a20} is known. The model has been tested, based on the experimental reports by several researchers, for a range of superficial gas flow rates (*0.08 m/min for the pilot scale tests as in Study Cases 1 and 2 to 0.18 m/min for the full-scale shop tests as in Case Study No. 3*), with the shop tests tank depths ranging from 3 m (10 ft) to 7.6 m (25 ft) under identical diffuser placement density. The model should be valid for these ranges of gas flow rates and tank heights. Therefore, the standard baseline $(K_La_0)_{20}$ determined from a single test tank is a valuable parameter that can be used to predict the $(K_La)_{20}$ value for any other test variables and gas flowrate (or height-averaged superficial gas velocity, U_g) by using the proposed model equations, provided that the tank horizontal cross-sectional area remains constant and uniform as the bubbles rise to the surface.

TABLE 5.9 Summary of the test results for standardized specific baseline mass transfer coefficient $(K_La_0)/U_g^q$

Case No.	Gas Type	Diffuser Type	Major Variables	Sp. K_La_{020}	Exponent q
1	Air	140-micron diffuser	Pressure, Q_a, gas composition	0.840	0.75
	Oxygen	140-micron diffuser	Pressure, Q_a, gas composition	1.511	0.92
2	Air	Slitted ADS tubing	Height, Q_a	0.994	0.71
3	Air	FMC fine pore tubing	Temperature, height, Q_a	0.861[*]	0.82
	Air	Norton fine dome	Temperature, height, Q_a	1.305	0.80
	Air	Pentech Jet	Temperature, height, Q_a	0.716	0.73

* Note: This calculated value was obtained from Chapter 3.

All the test results from the three experimenters are summarized in Table 5.9 below. The effective depth 'd_e' can be determined by solving a set of simultaneous equations but, in the absence of more complete data, 'e' can be assumed to be between 0.4 to 0.5 (Eckenfelder 1970) for ordinary air feed gas which has an oxygen mole fraction of 21%. For HPO, e can approach unity due to the higher driving force at bubble release as can be seen in Table 5.5 in Case Study No. 1. Yunt et al. (1988a) experiments, which were carried out in large tanks that resemble full-scale, further demonstrate that simulation and translation from one test to another is possible, with an error of not more than 3% in the estimation of K_La. The examples provided in this paper proved that the concept of a constant baseline for an aeration equipment is true for the range of gas flows and water depths tested. However, the tests were carried out with the tank horizontal cross-sectional area remaining constant and uniform as the bubbles rise to the surface for each test. Further testing is required on whether the same is true for tanks of different cross-sections for the same aerator, i.e. whether pilot test results can be translated to shop tests and to full-scale. For example, it would be interesting to find out if the specific baseline determined from Case 1 or 2 can extrapolate to the same results on Case 3 using the same aerator. The comparison graph in Fig. 5.12 below shows that the Norton diffuser is obviously superior since it has the highest baseline. However, when comparing the Pentech Jet and the FMC fine pore diffusers, it can be concluded that there is little to choose between the two for gas flow rates below 4.3 m³/min, but the FMC would become more superior if the design gas flow is beyond this gas flow rate. (Care must be taken in selecting the test gas flow rates and tank geometry such that other effects would not render the simulation model invalid. DeMoyer et al. (2002) and Schierholz et al. (2006) have conducted experiments to show the effect of free surface transfer on diffused aeration systems, and it was shown that high surface-transfer coefficients exist above the bubble plumes, especially when the air discharge (Q_a) is large).

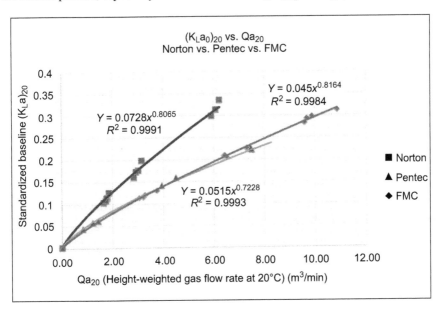

FIGURE 5.12 Baselines between Norton, Pentech jet, and FMC diffusers; data by Yunt et al. [1988a]

This good accuracy for estimating the baseline enables the production of rating curves for the aeration equipment under various operating conditions that include tank depths, pressures, temperatures as well as different gas flow rates. Therefore, given its importance, $(K_L a_0)_{20}$ should be expressed as an important parameter estimation to comply with the current standard. It can be used to evaluate the $K_L a$ in a full-size aeration tank along with the standard procedures covering the measurement of the oxygen transfer rate by any such submerged systems. The use of this parameter to determine $K_L a_p$ and whether the biological uptake rate R should be incorporated into the mass transfer equation as postulated in the Introduction, requires further investigation and is discussed in the next chapter in Chapter 6.

5.7 NOTATION (MAJOR SYMBOLS)

C^*_∞	oxygen saturation concentration in an aeration tank (mg/L)
$C^*_{\infty f}$	oxygen saturation concentration in an aeration tank under field conditions (mg/L)
$C^*_{\infty 20}$	oxygen saturation concentration in an aeration tank at 20°C (mg/L)
C_S	oxygen saturation concentration in an infinitesimally shallow tank; also known as oxygen solubility in water at test water temperature and barometric/headspace pressure; for open tank, also known as DO surface saturation concentration—handbook tabular values as a function of barometric pressure and water temperature (mg/L)
C	dissolved oxygen (DO) concentration in a fully mixed aeration tank (mg/L)
DO	dissolved oxygen in water (mg/L)
E	modulus of elasticity of water at atmospheric pressure, E/10^6 (kN/m²)
ρ	density of water (kg/m³)
σ	surface tension in contact with air (N/m)
R	microbial oxygen respiration rate (mg/L/(m³-hr))
R_0	specific gas constant for oxygen (kJ/(kg-K))
R^2	curve fitting correlation coefficient
$K_L a$	volumetric mass transfer coefficient also known as the apparent volumetric mass transfer coefficient (min⁻¹ or hr⁻¹)
$(K_L a)_{20}$	standard volumetric mass transfer coefficient (min⁻¹ or hr⁻¹)
$K_L a_f$	mass transfer coefficient as measured in the field (min⁻¹ or hr⁻¹)
$K_L a_0$	baseline mass transfer coefficient, equivalent to that in an infinitesimally shallow tank with no gas side oxygen depletion (min⁻¹ or hr⁻¹)
$(K_L a_0)_{20}$	standard baseline mass transfer coefficient, at zero gas side oxygen depletion (min⁻¹ or hr⁻¹)
z	depth of water at any point in the tank measured from bottom (m)
Z_d	submergence depth of the diffuser plant in an aeration tank (m)
Z_e	equilibrium depth at saturation measured from bottom (m)

P_a, P_b	atmospheric pressure or barometric pressure at time of testing (kPa; Pa)
P_e	equilibrium pressure of the bulk liquid of an aeration tank defined such that: $P_e = P_a + r_w d_e - P_{vt}$, where P_{vt} is the vapor pressure and r_w is the specific weight of water in kN/m^3 or N/m^3 (kPa; Pa)
d_e	effective saturation depth at infinite time (m)
e	effective depth ratio ($e = d_e/Z_d$)
Y_e	oxygen mole fraction at the effective saturation depth at infinite time
Y_0, Y_d	initial oxygen mole fraction at diffuser depth, Z_d, also equal to exit gas mole fraction at saturation of the bulk liquid in the aeration tank, $Y_0 = 0.2095$ for air aeration, $Y_0 = 0.80$ for HPO (high purity oxygen) aeration
H	Henry's Law constant (mg/L/Pa) defined such that: $C_\infty^* = HY_e P_e$ or $C_S = HY_0 P_a$
Y_{ex}	exit gas or the off-gas oxygen mole fraction at any time
y	oxygen mole fraction at any time and space in an aeration tank defined by an oxygen mole fraction variation curve
Q_a	height-averaged volumetric air flow rate (m^3/min or m^3/hr)
Q_{a20}	height-averaged volumetric air flow rate at 20°C (m^3/min or m^3/hr)
Q_S, AFR	gas (air) flow rate at standard conditions (20°C for US practice and 0°C for European practice), in (std ft^3/min or Nm^3/hr)
S	cross-sectional area of aeration tank (m^2)
U_g	superficial gas velocity given by Q_a/S (m/min; m/hr)
sp. $K_L a_0$	specific baseline mass transfer coefficient, $K_L a_0/U_g^q$ also expressed as $K_L a_0/Q_a^q$
sp. $(K_L a_0)_{20}$	standard specific baseline mass transfer coefficient, $K_L a_0/U_g^q$ also expressed as $(K_L a_0)_{20}/Q_a^q$
V	volume of aeration tank given by $S \cdot Z_d$ (m^3)
T_P	gas temperature at the point of flow measurement, Kelvin, assumed to be equal to the water temperature
P_p and P_b	the corresponding gas pressure and the barometric/headspace pressure, respectively, to T_P (Pa)
T	test water temperature in degree Celsius or in Kelvin
T_S	standard air temperature 20°C or 293 K
P_S	standard air pressure of 1.00 atm (101.3 kPa), or overhead pressure
P_d	total pressure at diffuser depth (kPa)
n, m	calibration factors for the Lee-Baillod model equation for the oxygen mole fraction variation curve
ϕ, x	system variables for Eq. 5.2 and Eq. 5.3 [$x = HR_0 T/U_g$; $\phi = x.(1 - e)$]
$\theta \cdot \tau \cdot \beta \cdot \Omega$	temperature, solubility, salinity, pressure correction factors as defined by the standard [ASCE 2007]

REFERENCES

Ahlert, R.C. 1997. Process dynamics in environmental systems. Edited by Walter, J.W. Jr., Francis, A.D. 1-943. New York: Wiley. https://doi.org/10.1002/ep.3300160107.

American Society of Civil Engineers. (1997). Standard guidelines for in-process oxygen transfer testing, 18-96. Reston, VA.

American Society of Civil Engineers. (2007). Standard 2-06: Measurement of oxygen transfer in clean water. Reston, VA.

Ashley Ken, I., Mavinic Donald, S., Hall, K.J. (first name n/a). (2009). Effect of orifice diameter, depth of air injection, and air flow rate on oxygen transfer in a pilot-scale, full lift, hypolimnetic aerator. Canadian Journal of Civil Engineering 36: 137-147, NRC Research Press.

Baillod, C.R. (1979). Review of oxygen transfer model refinements and data interpretation. Proc., Workshop toward an Oxygen Transfer Standard, U.S. EPA/600-9-78-021, W.C. Boyle, ed., U.S. EPA, Cincinnati, 17-26.

Barber, Tyler W. (2014). Superoxygenation: analysis of oxygen transfer design parameters using high purity oxygen and a pressurized aeration column. University of British Columbia.

Boon, Arthur G. (1979). Oxygen transfer in the activated-sludge process. Water Research Centre, Stevenage Laboratory, U.K.

Daniil, E.I., Gulliver, J.S. (1988). Temperature dependence of liquid film coefficient for gas transfer. Journal of Environmental Engineering, Oct 1988, 114(5): 1224-1229.

DeMoyer, Connie D., Schierholz, Erica L., Gulliver, John S., Wilhelms, Steven C. (2002). Impact of bubble and free surface oxygen transfer on diffused aeration systems. Water Research 37(8): 1890-1904.

Eckenfelder, Wesley W., Jr. (1966). Industrial Water Pollution Control, McGraw-Hill, New York.

Eckenfelder, Wesley W. (1970). Water Pollution Control: Experimental Procedures for Process Design. Pemberton Press, Austin.

EPA/600/S2-88/022. (1988). Project summary: aeration equipment evaluation. Phase I: Clean water test results. Water Engineering Research Laboratory, Cincinnati, OH.

Houck, D.H., Boon, Arthur G. (1980). Survey and evaluation of fine bubble dome diffuser aeration equipment. EPA/MERL Grant No. R806990.

Hwang, Hyung J., Stenstrom, Michael K. (1985). Evaluation of fine-bubble alpha factors in near full-scale equipment. Journal – Water Pollution Control Federation 57(12): 1142-1151.

Jiang, Pang, Stenstrom, Michael K. (2012). Oxygen transfer parameter estimation: impact of methodology. Journal of Environmental Engineering 138(2):137-142.

Kadlec, R.H., Reddy, K.R. (2001). Temperature effects in treatment wetlands. Water Environment Research 73(5): 543-557.

Lee, Johnny. (2017). Development of a model to determine mass transfer coefficient and oxygen solubility in bioreactors. Heliyon 3(2): e00248.

Lee, Johnny. (2018). Development of a model to determine the baseline mass transfer coefficient in bioreactors (Aeration Tanks). Water Environment Research 90(12): 2126-2140.

McGinnis, D.F., Little, J.C. (2002). Predicting diffused-bubble oxygen transfer rate using the discrete-bubble model. Water Research 36(18): 4627-4635.

McWhirter, John R., Hutter, Joseph C. (1989). Improved oxygen mass transfer modeling for diffused/subsurface aeration systems. AIChE J. 35(9): 1527-1534. https://doi.org/10.1002/aic.690350913.

Rosso, D., Stenstrom, M. (2006). Alpha Factors in Full-scale wastewater aeration systems. Water Environment Foundation. WEFTEC 06.

Schierholz, Erica L., Gulliver, John S., Wilhelms, Steven C., Henneman, Heather E. (2006). Gas transfer from air diffusers. Water Research 40(5):1018-1026.

Schroeder, E.D. (1977). Water and Wastewater Treatment. McGraw-Hill, Tokyo.

Stenstrom, Michael K. (2001). Oxygen transfer report: clean water testing (in accordance with latest ASCE standards) for Air Diffusion Systems submerged fine bubble diffusers on March 7, 8, 9, & 10 in 2001. http://www.aqua-sierra.com/wp-content/uploads/ads-full-oxygen-report.pdf.

Vogelaar, J.C.T., KLapwijk, A., Van Lier, J.B., Rulkens, W.H. (2000). Temperature effects on the oxygen transfer rate between 20 and 55 C. Water Research 34(3): 1037-1041.

Yunt, Fred W., Hancuff, Tim O., Brenner, Richard C. (1988a). Aeration equipment evaluation. Phase 1: Clean water test results. Los Angeles County Sanitation District, Los Angeles, CA. Municipal Environmental Research Laboratory Office of Research and Development, U.S. EPA, Cincinnati, OH.

Yunt, Fred W., Hancuff, Tim O. (1988b). EPA/600/S2-88/022. Project summary: aeration equipment evaluation. Phase I: Clean water test results. Water Engineering Research Laboratory, Cincinnati, OH.

Zhou, Xiaohong, Wu, Yuanyuan, Shi, Hanchang, Song, Yanqing. (2013). Evaluation of oxygen transfer parameters of fine-bubble aeration system in plug flow aeration tank of wastewater treatment plant. Journal of Environmental Sciences 25(2): 295-301. ISSN 1001-0742. https://doi.org/10.1016/S1001-0742(12)60062-X.

6

Is Oxygen Transfer Rate (OTR) in Submerged Bubble Aeration Affected by the Oxygen Uptake Rate (OUR)?

6.0 INTRODUCTION

Although the oxygen transfer rate and the oxygen uptake rate are two sides of the same coin, i.e. OTR = OUR, where the accumulation term in the bulk liquid is included in the OUR, the amount of evidence pointing to errors in the estimating of the oxygen mass transfer coefficient ($K_L a_f$) for diffused aeration is overwhelming (the subscript for $K_L a$ denotes mass transfer coefficient in the field), such as described in McCarthy (1982) where in section 3 under "Methods of Aeration Equipment Testing", it was stated that: "The need to accurately correlate clean water and wastewater test results... has been recognized by the U.S. Environmental Protection Agency (EPA) as an important area of research." Some literature has offered explanations, such as that the response time of probes caused the errors, or that the OUR measurement technique is faulty, etc., but none of these explanations are convincing enough to explain the non-correlation between the clean water and the wastewater mass transfer coefficients. In fact, as early as 1979, experiences have indicated that OUR under 60 mg/L/hr can be measured with a minimum of error (McKinney and Stukenberg 1979). As for the probes, the lag times for modern fast-response probes have been drastically reduced (Baquero-Rodríguez and Lara-Borrero 2016, Baquero-Rodríguez et al. 2018) (ASCE 2007). No one has ever considered that the equations for the mass balancing might be incomplete.

The purpose of this manuscript is to address the anomaly in $K_L a_f$ estimation ($K_L a$ vs. $K_L a_f$) by re-examining the equations as given in Section 2 and Section 3 of the ASCE Standard Guidelines for In-Process Oxygen Transfer Testing (ASCE 1997). For simplicity, the following arguments pertain to a completely-stirred batch process only, with a tank/vessel volume of unity. The author postulates that inconsistency in the evaluation of $K_L a_f$ between the Non-steady State Methods (*Changing Power levels (CPL) or Hydrogen Peroxide Addition (HPA)*) and the Steady State Methods (*Oxygen Uptake Rate (OUR) or Off-gas (OG)*) for submerged diffused aeration is caused by the different gas depletion rates (gas depletion is defined as the difference between the oxygen content of the feed and exit gas due to the loss of oxygen partial pressure as the bubbles rise to the surface) during testing between the two broad categories of methods under the same mass gas flow rate and substrate loading conditions prior to testing. This difference in gas depletion rates (*gdp*) must be accounted for in the mass balancing equations. In in-process water, care must be taken to ensure that the parameter $K_L a_f$ is *not* a function of dissolved oxygen concentration. This dependency

116

can occur where air is injected through diffusers on the bottom of activated sludge tanks or fermentation bioreactor vessels, where rising air bubbles are significantly depleted of oxygen as they ascend to the water surface [CEE 453, 2003] [Rosso and Stenstrom 2006a]. The extent of oxygen depletion is a function of the oxygen concentration in the activated sludge mixed liquor.

K_La, by definition, is the product of the liquid film coefficient K_L and the interfacial area of the gas-liquid interface. (The theory of oxygen transfer is given in Chapter 4, and is based on a mole fraction variation curve as shown in Fig. 6.1.) Why would K_La be dependent on anything else? Although the oxygen transfer rate is affected by the microbial cells of the activated sludge, K_La itself should not, since, in principle, it is not physically or chemically or biologically connected to the microbial interactions of any live microorganisms. The availability of oxygen has nothing to do with this parameter, although the oxygen availability in terms of the dissolved oxygen concentration may be a critical parameter for ethanol production (in fermentation) or biomass production.

A study with a control chemostat in parallel, but with the cells killed off either by UV or by chemicals such as sulfamic acid-copper sulfate, may illustrate this point. The characteristics of the liquid interface would indeed affect diffusivity and hence K_L, and the interfacial area may be affected by many factors such as the air supply system. Looking at all the literature, there is no evidence that K_La would be affected by live organisms. The non-steady state method (gassing-in method in the language of fermentation literature) should give a more reliable value of K_La, since it is free from any interference from any live organisms if first destroyed, and it should compare well with the steady-state method. Unfortunately, they usually differ by as much as 50%.

Shraddha and Narang (2018) have postulated that one must assume that the K_La measured in cell-free medium persists in the presence of growing cells. However, according to Shraddha and Narang (2018), this assumption is not tenable in their experiments because the rheology of the culture changes with the operating conditions — the culture is viscous and foams during dual limited (substrate-limited and oxygen-limited) growth. However, when comparing two cultures, one with living cells and the other without, the media can be made the same at a certain fixed set of operating conditions, but with the control devoid of the cells. The control can be used to measure K_La using the non-steady state method, while the other one's K_La can be determined by the steady-state method. In principle, both values should be the same. Garcia et al.'s experiments [Garcia-Ochoa and Gomez 2009, Garcia-Ochoa et al. 2010] [Santos et al. 2006] using *Xanthomonas compestris* culture in one test, and *rhodococcus erythropolis* culture in another, was illustrated by the author in detail (Fig. 6.5, Fig. 6.6 and Fig. 6.7). These experiments showed a discrepancy between the steady-state and non-steady state tests of around 50%. If one modifies the transfer equation to include the gas depletion (*gdp*) effect, as the author had tried with Garcia et al.'s data, the author found that the same value of K_La is arrived at in both methods. The gas depletion rate is no more than the respiration rate R, and the transfer equation becomes $dC/dt = K_La (C_S - C) - R\text{-}gdp$ which, when equating R with gdp, gives $dC/dt = K_La (C_S - C) - 2R$.

Therefore, the author suspects that this discrepancy in the conventional model is due to the mass balance equations missing some important factors, such as the effect of gas-phase gas depletion during aeration that is different between the two cases (steady-state versus non-steady state), making the former 50% less than the latter. If this is accounted for, the two values should be similar. Unfortunately, few people have done such experiments. The nearest such experiments are given in Garcia-Ochoa et al. (2010) as shown in Fig. 6.5, Fig. 6.6 and Fig. 6.7.

In the author's opinion, only when the oxygen transfer coefficient is accurately measured can the oxygen uptake rate be accurately determined, since, in a steady state, the OTR and the OUR_f are two sides of the same coin, and the former is dependent on K_La. There is no accumulation term in a steady state. Therefore, at steady state, the oxygen uptake rate OUR_f is the microbial respiration rate R. If K_La_f is not estimated correctly, then these two terms (OTR, OUR_f) will not balance each other, with the usual culprit being the OUR_f as mentioned in the first paragraph. This happens when a method such as the off-gas steady-state method is chosen to estimate K_La_f, leading to an erroneous estimation of the OTR_f. On the other hand, if K_La_f is measured correctly, such as by the non-steady state method, or by Garcia's gassing-in method, then any independent separate measurement of the OUR_f will not give the same K_La_f, leading to doubts about the OUR_f method of measurement and/or the steady-state method itself [ASCE 1997].

In an aeration tank of a wastewater treatment plant, when a steady state is reached (i.e. the oxygen supply meets the oxygen demand from the biological system), the mole fraction of oxygen in the gas phase would decrease as the depth decreases, so that the exit gas has a smaller mole fraction than the feed gas (see Fig. 6.1 in Section 6.1.2). There is evidence that the gas depletion rate or the oxygen transfer rate is affected by any biochemical reactions such as the respiration rate of any microorganisms occurring within the liquid, as shown in Fig. 6.3 and Fig. 6.4. The hypothesis presented in this manuscript is that, for the same gas supply rate, the effect of such reactions is a negative impact on gas depletion, so that the higher the reaction rate, the smaller is the gas depletion rate, and therefore less gas will be transported or transferred into the liquid under aeration. In mathematical terms, $F_1 - F_2 = R$, where F_1 is the gas depletion rate unaffected by any biochemical reactions; F_2 is the gas depletion rate in the presence of biochemical reactions in the liquid, and R is the reaction rate or the microbial respiration rate or the microbial oxygen uptake rate (steady-state OUR_f). This effect of changes in the gas depletion rate with respect to changes in the mixed liquor suspended solids (MLSS) under a constant gas flow rate is illustrated by Hu (2006), as shown in Fig. 6.3.

This chapter presents mass balance equations that would include the gas depletion effect, so that the testing methods give consistent results, and that the measured mass transfer coefficients in the field can be related to clean water K_La. Based on data on tests extracted from literature, the proposed *revised* equations for the American Society of Civil Engineers (ASCE 1997) testing methods are shown to result in a consistent estimation of the mass transfer coefficient (K_La_f), where previously the estimation among the methods generally had a discrepancy of around 40 to 50%.

6.1 THEORY

6.1.1 Relationship Between $K_L a$ and Water Characteristics

In oxygen transfer, Lewis and Whitman (1924) advanced the two-film theory as a classical theory of aeration. Using this theory, most models simulate the movement of gas into the water, but not the other way around. In fact, gas transfer is a two-way street because the gas dissolves into the water as well as flows back to the gas stream from the water (Baillod 1979, Jiang and Stenstrom 2012). To make a truly accurate model, one must simulate the dynamic movement.

After so many years of research since the inception of the activated sludge technology, the author believes that it is a universally established acceptance that the parameter $K_L a$, whether it be for clean water or wastewater, or mixed liquor, is a function ONLY of the physical characteristics of the water involved, so long as the external variables such as temperature, pressure, tank geometry, diffuser plant, solute gas, gas flow rate, turbulence, etc. are not changed. If this sole dependence is not accepted, it will be necessary to thoroughly review the fundamentals of oxygen transfer, the two-film theory, etc., and re-visit all the researches carried out so far in the literature world on this topic. This accepted understanding has led to the standard oxygen transfer equation given by ASCE 2-06 [2007], Metcalf and Eddy [1985], etc.:

$$\frac{dC}{dt} = K_L a (C^* - C) \tag{6.1}$$

where, although the meanings and definitions of the symbols may change with respect to the applications being applied to, the general form of the equation appears to hold for any systems of oxygen transfer in liquid water under any conditions, no matter whether it is clean water or dirty water. C^* can have different meanings within the context of each application, but it always pertains to an equilibrium concentration value.

For the main issue that the manuscript aims at solving (the gap between gas transfer rate under clean water and wastewater conditions), conventional method is that the consumption of oxygen due to biological reactions is dealt with using a coupled mass balance equation, while the mass transfer equation is well maintained in its current, widely accepted form. In other words, the right-hand side of equation (Eq. 6.1) can still be used to describe the mass transfer rate; however, the change in oxygen concentration in the liquid phase will be influenced by mass transfer and biological consumption R. The salient equation for this scenario is:

$$\frac{dC}{dt} = K_L a_f (C^*_{\infty f} - C) - R \tag{6.2}$$

This equation has recognized that R is an additive quantity and not a scalar quantity associated with $K_L a$ but is still flawed because, granted that such a unique function of variability of $K_L a$ is an accepted fact, the water characteristics may be changed by outside factors, such as the quantity and character of suspended solids in the water. Suspended solids concentrations, perhaps along with other constituents, change the viscosity and density of the liquid and hence affect $K_L a$. Lee (2017) has shown, using water temperature as the independent variable, that $K_L a$ in fact is related to the above water properties, as well as surface tension. When comparing wastewater mass transfer coefficients (K_La_f) with these altered characteristics to the corresponding clean water coefficients (K_La), the K_La value usually becomes slightly smaller (Tchobanoglous et al. 2003). In England, the Water Research centre (WRc) had used detergent added to clean water to mimic municipal wastewater, so that the measured K_La would be representative of the field K_La without having to measure the *in-situ* K_La in a treatment plant [Boon 1979] [ASCE 2007].

Now, if the conventional model ASCE 18-96 equation 2 [ASCE 1997] is correct, so long as the water characteristics remain constant, the measured K_La should be a constant, but it is not, since the value of K_La in the field is highly variable with respect to the microbial population that can be represented by the oxygen uptake rate R. For the same oxygen supply rate, the higher the value of R, the lower is the value of the measured K_La. Hwang and Stenstrom (1985) plotted the 'mass transfer coefficient' versus 'oxygen uptake rate' in decaying OUR tests and showed a linear declining trend with increasing R. Why?

There are only two possible explanations: one is that K_La is in fact a function of R. The other explanation is that the mass balancing equation is wrong or incomplete for diffused aeration because it hasn't taken into account the changes in the gas depletions due to microbial respiration. (Gas depletion is defined as the difference between the oxygen content of the feed and exit gas due to the loss of oxygen partial pressure as the bubbles rise to the surface). If the first explanation is correct, logically it overturns the established concept that K_La is only dependent on the water characteristics and properties. However, it can be argued that the nature of the water may change due to the presence of microorganisms. Indeed, Hwang and Stenstrom (1985) showed in another graph that as R increases, the surface tension (as measured by the Du Nouy ring method) decreases. But since previous plotting has showed a linear declining trend of K_La in relation to R, it can be argued that the altered water characteristics due to the microbes exert a resistance to oxygen transfer and therefore this resistance decreases the K_La. This then conforms to the concept that K_La is only a function of water characteristics and not a function of R.

But where does this additional alteration of the characteristics come from? Why would the microbial respiration alter the surface tension of the water in question? This is obviously something for future research. But for now, it can be argued that the presence of the living microbes or the biochemical reactions associated with their metabolism provide the alteration, and so the change in the corresponding gas depletion rate that changes the K_La must come from the microbial respiration itself. There is no other factor that may have caused that change. However, unlike water characteristics, the effect on the gas depletion rate is additive, not associative. While one can use a partial factor, alpha (α), on K_La to account for the changes in water characteristics, whose property is intensive (i.e. not dependent on scale as long as the fluid is well-mixed), one cannot do the same for the gas depletion rate which is reactive and consecutive to the respiration rate. Hence, changes in gas depletion rate due to the microbes is an extensive property (i.e. the oxygen uptake rate of the microbes can be changed substantially without effecting a substantial change in the water characteristics and properties, even though some changes are inevitable, such as surface tension). The consequence of R is therefore mostly additive in the mass balance equation.

If the second explanation is correct, and since K_La_f should be constant when the water characteristics are constant, then the oxygen transfer rate cannot be just given by the ASCE equation 2, which has not accounted for that additional change in the gas depletion rate due to the microbial respiration that changes the surface tension. Based on the above argument, the first explanation is not entirely incorrect. However, the variation in water characteristics due to the microbes is much less than the variation in the gdp because R is highly variable. This is especially true when the situation is close to endogenous where the DO is very low, and the gdp is at a minimum due to the high biological activity and the high resultant resistance. At the same time, it is also at a maximum because of the high driving force. The conventional model adopts a holistic approach, in which K_La_f is corrected by alpha (α) associated with the clean water K_La that does not adequately account for its changes due to the gas depletion effect. Using alpha purely to correct for the water characteristics would give a much more consistent value of K_La_f. This manuscript introduces the hypothetical concept that the difference in the respective gas depletion rates is precisely equivalent to the microbial respiration rate R. If this is true, it has explained why, in the proposed equation by the author, the $2R$ is required instead of just a single R which is only correct for surface aerators where the phenomenon of gas depletion doesnot exist.

6.1.2 Le Chatelier's Principle Applied to Gas Transfer

Conceptually, before reaching the saturation state in a non-steady state test, since the oxygen concentration in the water is less than would be dictated by the oxygen content of the bubble, Le Chatelier's principle requires that the process in the context of a bubble containing oxygen and rising through water with a dissolved-oxygen deficit, relative to the composition of the bubble, would seek an equilibrium via the net transfer of oxygen from the bubble to the water [Mott 2013]. In this scenario, even for the ultimate steady-state, oxygen goes in and out of the gas stream depending on position and time of the bubble of the unsteady state test. In clean water, one can view the mass balances as having two sinks—one by diffusion into water, and the other by diffusion from water back to the gas stream which serves as the other sink. Whichever is the greater depends on the driving force one way or the other. At system equilibrium, these two rates are the same at the equilibrium point of the bulk liquid, the equilibrium point being defined by the effective depth 'd_e' in ASCE 2-06 [ASCE 2007]. At steady state, the entire system is then in a dynamic equilibrium, with gas depletion at the lower half of the tank below the 'd_e' level, and gas absorption back to the gas phase above d_e-the two movements balancing each other out. Therefore, the general understanding that" *the overall mass transfer coefficient "K_La" incorporates the mass transfer through the gaseous and liquid films at equilibrium"* is applicable to clean water only, if equilibrium is being defined as a state where the gas flowing into the *bulk liquid* equals the gas flowing out of the *bulk liquid*. In the author's opinion, only clean water tests can achieve equilibrium where the potential to transfer (fugacity) is fully utilized. At equilibrium, which is also steady state, the inlet feed gas mass flow rate will be equal to the exit gas mass flow rate. On the contrary, in in-process wastewater, only steady state can be achieved, as the fugacity may not be fully utilized. When the DO changes, or the OUR changes, the potential to transfer may change accordingly. This can be understood by examining the gas side oxygen depletion curve, where the exit gas oxygen mole fraction is lower than the feed gas mole fraction at steady state for process water. The steady state DO concentration (C_R) is only an "apparent" saturation concentration that is not stable as opposed to the saturation concentration in clean water when steady state is achieved. However, the effect of gas depletion must be considered in both cases, as this manuscript explains below. Equilibrium means that the fugacity of the oxygen in the gas phase is equal to the fugacity of the oxygen in the liquid phase and LeChatelier's principle applies to equilibrium.

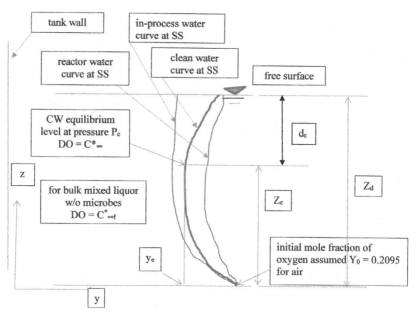

FIGURE 6.1 Oxygen mole fraction curves at saturation for both in-process water and clean water based on the Lee-Baillod model (e = equilibrium) [Lee 2018]

If the system is at equilibrium, then it is at steady state, as shown in Figure 6.1 at clean water saturation (also shown in Fig. 3.1 of Chapter 3). As mentioned, the mole fraction variation curve for any in-process water will not reach equilibrium even at steady state (SS), as shown by the other curve on the left of the clean water curve. If the aqueous solution has no microbes in it, the curve would be like the one on the right of the clean water curve, but the equilibrium mole fraction would be slightly less due to any additional resistance in the liquid. The standard mass transfer model for a bulkliquid aeration under constant gas flow rate has been theoretically derived in Chapter 4, given by Eq. 4.1 repeated herewith as Eq. 6.3:

$$\frac{dC}{dt} = K_L a(C_\infty^* - C) \qquad (6.3)$$

For the general case, the equation for K_1 in Eq. 4.44 in the previous Chapter 4 (K_1 is defined in Chapter 3 Eq. 3.3 and in Chapter 4 Eq. 4.44) can be modified to

(Eq. 6.4) below [Lee 2018], and the generalized Lee-Baillod equation (Eq. 6.5) can be subjected to mathematical integration to yield Eq. 6.6 and Eq. 6.7, just like the previous case for the CBVM (constant bubble volume model) [Lee 2018]. All the resulting equations that lend themselves to five simultaneous equations for solving the unknown parameters (n, m, K_La_0, y_e, Z_e) are reiterated and summarized below:

$$K_La = \frac{[1-\exp(-K_La_0x(1-e)Z_d)]}{x(1-e)Z_d} \tag{6.4}$$

$$y = \frac{C}{nHP} + \left(\frac{Y_0P_d}{P} - \frac{C}{nHP}\right)\exp(-xK_La_0 \cdot mz) \tag{6.5}$$

$$C_\infty^* = nH(0.2095)\frac{P_a - P_d\exp(-mx \cdot K_La_0 \cdot Z_d)}{1 - \exp(-mx \cdot K_La_0 \cdot Z_d)} \tag{6.6}$$

$$K_La = \frac{1 - \exp(-mx \cdot K_La_0 \cdot Z_d)}{nmx \cdot Z_d} + \frac{(n-1)K_La_0}{n} \tag{6.7}$$

$$Z_e = \frac{1}{mxK_La_0}\left\{\ln\left(Pe\frac{mxK_La_0}{nr_w}\right) + \ln\left(\frac{nHY_0P_d}{C_\infty^*} - 1\right)\right\} \tag{6.8}$$

(where $x = HR_0T/U_g$, where U_g is the height-averaged superficial gas velocity; R_0 is the specific gas constant of oxygen (note: a different symbol is used to distinguish it from the respiration rate R); T is the water temperature; e is the effective depth ratio $e = d_e/Z_d$). Hence, the basic transfer equation for the non-steady state clean water test as given by Eq. 6.3 is proven for the general case (non-constant bubble volume) as well, where K_La and C_∞^* are obtainable by solving the above set of equations when the baseline K_La_0 is known. Based on the above derivation, by the principle of mathematical induction, it can be argued that, for very shallow tank $(Z_d \approx 0)$, the basic transfer equation is again applicable. Hence, the following equation would apply:

$$\frac{dC}{dt} = K_La_0(C_S - C) \tag{6.9}$$

where C_S is the handbook solubility value at the atmospheric pressure and water temperature at testing. Comparing Eq. (6.1) with Eq. (6.9), the two mass transfer coefficients are not the same, since the former has incorporated the effect of gas depletion as seen in the derivation (see Chapter 4), whereas in the latter equation, gas depletion is non-existent because of the zero depth. However, for tank aeration with gas depletion, Eq. (6.3) can be modified to:

$$\frac{dC}{dt} = K_La_0(C_{\infty 0}^* - C) - gdp_{cw} \tag{6.10}$$

where K_La_0 is as calculated by Eq. 6.4 from a known value of K_La. The parameter $C_{\infty 0}^*$ is the saturation concentration that would have existed without the gas depletion (note that $C_{\infty 0}^{*}$ is not C_S), and gdp_{cw} is the overall gas depletion rate during a clean water test. This equation is based on the *Principle of Superposition* in physics where the transfer rate is given by the vector sum of the transfer rate as if gdp (gas depletion rate) does not exist, and the actual gas depletion rate which is a negative quantity. $C_{\infty 0}^{*}$ cannot be the same as C_S because the latter is the oxygen solubility under the condition of 1 atmosphere pressure only, while $C_{\infty 0}^{*}$ should correspond to the saturation concentration of the bulk liquid under the bulk liquid equilibrium pressure, but without the gas depletion (this of course cannot happen, since without gas depletion there can be no oxygen transfer). The hypothetical $C_{\infty 0}^{*}$ must therefore be greater than C_{∞}^{*} which in turn is greater than C_S since the former corresponds to a pressure of P_e while the latter corresponds to the free surface pressure P_a. This method of reasoning allows solving for the transfer from the baseline mass transfer coefficient as shown in Eq. 6.10. Since $K_L a$ is a function of gas depletion, and since every test tank may have different water depths and different environmental conditions, their gas depletion rates are not the same; hence, they cannot be compared without a baseline [Lee 2018]. Furthermore, by introducing the term gdp_{cw}, the oxygen transfer rate based on the fundamental gas transfer mechanism (the two-film theory) can be separated from the effects of gas depletions on $K_L a$. This gas depletion rate cannot be determined experimentally, since gdp varies with time throughout the test. Jiang and Stenstrom (2012) have demonstrated the varying nature of the exit gas during a non-steady state clean water test. Therefore, the only equation that can be used to estimate the parameters is Eq. 6.1 (where $C^{*} = C_{\infty}^{*}$ in the transfer equation) giving Eq. 6.3 above:

$$\frac{dC}{dt} = K_L a (C_{\infty}^{*} - C)$$

Eq. 6.3 is essentially equivalent to Eq. 6.10 but expressed differently ($K_L a$ vs. $K_L a_0$). Therefore, by the same token using the *Principle of Superposition*, for in-process water without any microbes, Eq. 6.10 would become Eq. 6.11:

$$\frac{dC}{dt} = K_L a_{0f} (C_{\infty 0f}^{*} - C) - gdp_{ww} \qquad (6.11)$$

giving, in the presence of microbes:

$$\frac{dC}{dt} = K_L a_{0f} (C_{\infty 0f}^{*} - C) - gdp_{ww} - gdp_f - R \qquad (6.12)$$

or when expressed differently using the measurable parameters:

$$\frac{dC}{dt} = K_L a_f (C_{\infty f}^{*} - C) - gdp_f - R \qquad (6.13)$$

where gdp_f is the gas depletion rate due to the microbial respiration. Note that in this equation, when $dC/dt = 0$, gdp_f would be given by $K_L a_f (C_{\infty f}^{*} - C) - R$, where C becomes a constant, usually denoted by C_R as the apparent saturation concentration at steady state.

6.1.3   The Hypothesis of a Baseline $K_L a$ for Wastewater

When the system has reached a steady state in the presence of microbes, the gas depletion rate is a constant, and so it would be possible to calculate the microbial gdp by the same equation and by incorporating R as well when $dC/dt = 0$ and $C = C_R$. In the presence of microbes, the advocated hypothesis is that this gdp_f due to the microbes is the same as the reaction rate R and so $dC/dt = K_L a_f (C_f^* - c) - R$ $- R$, compared to clean water where the microbial $gdp = 0$. In other words, if F_1 is the gas depletion rate for clean water, and F_2 is the gas depletion rate in process water, then $F_1 - F_2 = R$. It should be noted that, as mentioned before, the basic mass transfer equation is universal, its general form given by Eq. (6.1). Therefore, in a non-steady state test for in-process water for a batch test, the transfer equation is given by:

$$\frac{dC}{dt} = K_L a_f (C_R - C) \tag{6.14}$$

where C_R is the "*apparent*" saturation concentration or the "*pseudo*" steady-state DO value in the test tank at the *in-situ* oxygen uptake rate, R. But the transfer equation is also given by $dC/dt = K_L a_f (C_{\infty f}^* - c) - R - R$. Equating the two gives,

$$K_L a_f (C_{\infty f}^* - C) - R - R = K_L a_f (C_R - C) \tag{6.15}$$

which gives:

$$K_L a_f = \frac{2R}{(C_{\infty f}^* - C_R)} \tag{6.16}$$

Note that in this equation, C is cancelled out, so that the above equation is valid for any value of C, at any state, so long as $dC/dt > 0$ and $C < C_R$. Most models did not simulate the gas phase, and so are missing this important element in their balancing equations. This $K_L a_f$ can then be related to the clean water $K_L a$ which serves as a baseline for extrapolating the clean water test results to wastewater.

ASCE Guidelines 18-96 [ASCE 1997] reported that "in an EPA Cooperative Agreement research program, side-by-side comparisons were made of process water oxygen transfer test procedures (Mueller and Boyle 1988). Based on these test results on the estimation of $K_L a_p$ it was concluded that steady-state testing using oxygen uptake rates, although the easiest procedure to conduct, is not recommended because it may significantly overestimate or underestimate the real oxygen transfer rate. Overestimates are detected in low DO systems. Underestimates appear to be caused by the presence of a readily available exogenous food source that is rapidly consumed, and, therefore, is not effectively measured (as uptake) in samples removed from the basin."

Mahendraker et al. (2005a) compared oxygen transfer test parameters from four testing methods in three activated sludge processes and found different kinds of discrepancies from the ASCE Guidelines. The concept of an additional resistance is advanced by Mahendraker (2003) with the advocacy of a net respiration flux, and Mahendraker et al. (2005b) who demonstrated the different mass transfer coefficients taking into account the gas depletion effect by the floc.

It is notable that the conclusion reached by Mahendraker et al. (2005a) about the methods is completely opposite to the ASCE Guidelines. In their paper, it was the non-steady method that was considered suspect, and the steady-state methods *Oxygen Uptake Rate (OUR) or Off-gas (OG)* were considered correct because of the close agreement between these two in their estimation of K_La_f (In fact, both these methods give an erroneous estimation of K_La_f but by the same amount, lead to the mistaken conclusion that they were better methods than the non-steady state method). The author postulates that all methods will be correct if the mass balance equation has included the effect of *gdp*, notwithstanding the various legitimate defects for each individual method of testing.

Paradoxically, Boyle and Campbell (1984) appear to agree with Mahendraker et al. (2005a) overall conclusion that the steady-state method is valid, as can be seen from table 6 of their report, where the OTE (oxygen transfer efficiencies) of the off-gas method and the steady-state method in testing on a municipal wastewater treatment plant are compared. Furthermore, they found that excellent agreement between the gas tracer method which is considered as a referee method (ASCE 1997), and the off-gas method was achieved in another experiment, therefore suggesting that the steady-state method is accurate as well. Unfortunately, unlike in Mahendraker et al. (2005a) experiment, the non-steady state method (NSS) was not compared. Had it been done, they would have found an anomaly in the K_La_f estimates as Mahendraker et al. (2005a) have found between the NSS method and the SS method. It should be noted in passing that during the development of the inert gas radiotracer procedure for oxygen transfer measurement, the ratio K_{Kr}/K_La, where K_{Kr} is the volumetric gas transfer rate coefficient for krypton-85, was determined experimentally in laboratory studies using surface aeration apparatus. The value obtained, 0.83, has been proven accurate for surface transfer systems only. In sub-surface aeration, the effect of gas depletion must be considered, as can be seen in Eq. 6.12 above. The gas depletion not only comes from the depth of aeration, but also from the microbial oxygen uptake rate R, which according to the hypothesis in this manuscript is equivalent to the attendant gas depletion rate coming from the resistance of biological floc, and is an associative-additive quantity in the oxygen transfer Standard Model. Therefore, the ratio K_{Kr}/K_La may not be 0.83 for a sub-surface system under process conditions. It is also important to note that the krypton method only gives K_La estimation, whereas the off-gas method gives estimation of OTE, necessitating calculation for the K_La from the OTR test results, based on the standard model. If the calculation has not included the gas depletion effect, the good match between the two methods is only a coincidence, since both methods have neglected to take into account gas depletion in their estimation of K_La_f. The same argument goes for the oxidation ditch tests carried out by Boyle et al. (1989).

6.1.4   The Application of the Baseline K_La for Wastewater

However, it appears that all these discrepancies can be explained by bearing in mind that in the equation $K_La_f = \alpha(K_La)$, where K_La is a baseline based on a clean water test, α represents a contamination partial factor dependent **only** on the water characteristics (α is around 0.8 for domestic wastewater usually).

To distinguish this ratio for the baseline case from the other case where α is directly measured from the field, it would be better to use a different symbol, such as

α' (Mahendraker et al. (2005b) used the symbol α_e to represent the same parameter). Based on this modified equation, since $dC/dt = OTR_f - R$, then $OTR_f = \alpha'K_La(C^* - c) - R$, which says OTR is affected negatively by the OUR. The higher the value of R, the lower is the transfer rate. This hypothesis concurs with Hwang and Stenstrom's [1985] findings. According to their finding, the degree of reduction of K_La_f due to the microbial respiration R is dependent on tank depth, and the air flow rate. Since the gas depletion rate as seen in the Lee-Baillod model [Lee 2018] is also dependent on tank depth and the air flow rate, it can be demonstrated that, by using the data from the literature [Mahendraker 2003], the microbial gas depletion rate and the respiration rate are the same. This makes the translation of the clean water baseline K_La to dirty water K_La_f mathematically possible (note: $V = 1$ in the discussion).

The main problem in modeling is the issue of scale. In a full-scale plant, many factors come into play that would affect the mass transfer coefficient, so that the parameter α becomes a variable. One such factor is the gas depletion rate in a diffused submerged bubble aeration system as mentioned before. However, the alpha' (α') factor, which is the ratio of mass transfer coefficients between dirty water and clean water, is an intensive property if the physical characteristics of the wastewater is constant, and so can be established by bench-scale experiments [Eckenfelder 1970]. Due to gas depletion, the author has previously developed an equation that would give a more realistic K_La_f value in full-scale, based on their different tank heights [Lee 2018]. The derived equation is given below as:

$$K_La_f = \frac{1 - \exp(-\Phi Z_d \cdot \alpha K_La_0)}{\Phi Z_d} \tag{6.17}$$

where K_La_0 is again a baseline in the context of eliminating gas depletion effects due to tank height, similar to Eq. (6.4), and $\alpha = \alpha'$.

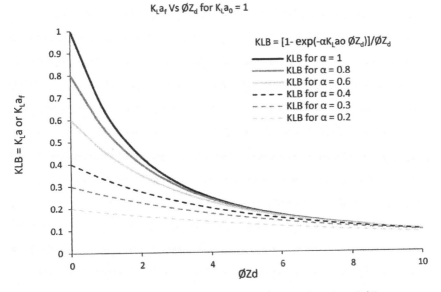

FIGURE 6.2 Apparent mass transfer coefficient vs. function of ϕZ_d

Figure 6.2 (reproduced from Fig. 3.10) shows the effect on the parameter estimation for different values of α, assuming $K_L a_0 = 1$. Notice that the value of Φ given by $\Phi = x \cdot (1 - e)$ may change when applied to wastewater (see section 7.2.2 in Chapter 7).

Equation (6.17) then accounts for the gas depletion effect and allows translation from the baseline to full-scale $K_L a_f$. Hence, for the mass balance as shown by the ASCE equation 2, for the case where continuous wastewater flow is absent, although the author agrees that the respiration rate (R) must equal transfer rate *minus* any accumulation rate that is occurring via a changing dissolved oxygen (DO) level, such that at steady state the OTR must equal the respiration rate R or the OUR, the author challenges the conventional thinking for the expression of the transfer rate using the transfer coefficient $K_L a_f$ that has not included gas depletion. The question presented in this manuscript is:

"In submerged aeration, should the oxygen transfer rate (OTR) be given by $K_L a_f \cdot (C^*_{\infty f} - C)$ or should it be $K_L a_f (C^*_{\infty f} - C) - R$?" Using the baseline $K_L a$ for clean water, it would appear that the latter is correct because of the different gas depletion rates between clean water (or non-respiring water) and in-process water where microbial cells are active with a respiration rate of R. Below is an example of calculating the baseline using a typical case [Baillod 1979]:

Suppose the following results are used/obtained by a clean water test:

$$Z_d = 3.05 \text{ m (10 ft)}$$
$$T = 19.5°C$$
$$H = 4.4248 \times 10^{-4} \text{ mg/L/(kPa)}$$
$$C_S = 9.2 \text{ mg/L}$$
$$C = 2 \text{ mg/L} \quad \text{(measured or calculated)}$$
$$Y_d = 0.2095$$
$$Y_{ex} = 0.19 \quad \text{(measured or calculated)}$$
$$\rho_a Q_a / A = 197 \text{ g/min-m}^2 \text{ or } 11.82 \text{ kg/hr/m}^2$$

(The product $\rho_a Q_a$ is the mass flow of air at standard conditions; A is the horizontal area of tank.

This gives Q_a per unit area as 0.1636 m³/min/m² when $\rho_a = 1.204$ kg/m³ at 20°C)

$$K_L a = 0.15 \text{ min}^{-1} \quad \text{(measured)}$$
$$C^*_\infty = 10.0 \text{ mg/L} \quad \text{(measured)}$$

From re-arranging Eq. 6.5, we have,

$$C = nHY_0 \frac{\left(\dfrac{Py}{Y_0} - P_d \exp(-xK_L a_0 \cdot mz)\right)}{(1 - \exp(-xK_L a_0 \cdot mz))} \tag{6.18}$$

At $z = Z_d$, $y = Y_{ex}$, $Y_0 = Y_d = .2095$, $P = P_a$, C can be calculated to be 2 mg/L, which confirms the measured value (note that if C was measured instead, the same equation could be used to calculate the exit gas mole fraction Y_{ex}). The calibration factors (n, m) and other variables, including the baseline $K_L a_0$ [Lee 2018], can be found by using the Excel Solver as shown in Table 6.1, assuming the exit gas mole fraction is now 0.1900. Assuming pseudo-steady state (i.e. the mass flow rate has only negligible changes during the transit from tank bottom to top), the gas depletion rate (*gdp*) is approximately given by:

$$gdp = \rho_a Q_a (Y_0 - Y_{ex}) \qquad (6.19)$$

Hence,

$$gdp = 1.204 \text{ kg/m}^3 \times 0.1636 \text{ m}^3/\text{min/m}^2 \ (.2095 - .1900) \times 60 \text{ min/hr} = .23 \text{ kg/hr/m}^2$$

On the other hand, the oxygen transfer rate OTR at $C = 2$ mg/L is also given by the liquid phase mass balance as:

$$OTR_{C=2} = K_L a \text{ min}^{-1} (C_\infty^* - 2) \times .001 \text{ kg/m}^3 \times 60 \text{ min/hr} \times 3 \text{ m} = 0.21 \text{ kg/hr/m}^2$$

which is close to the *gdp* as calculated by the gas phase mass balance.

TABLE 6.1 Calculation of the baseline mass transfer coefficient [Lee 2018]

Environmental Data								
	Water temp.	$T =$	19.5	°C				
	atm press.	$P_a =$	101325	N/m^2				
	Tank area	$S =$	1	m^2		**Error Analysis**		
Calc. Variables						Eq. I =	1.79E-05	3.20E-10
	Diff press.	$P_d =$	128359	N/m^2		Eq. II =	2.98E-06	8.88E-12
	$x =$	$x =$	0.2058	min/m		Eq. III =	4.25E-07	1.81E-13
	HR'ST/Q							
Solver Soln.	Baseline $K_L a$	$K_L a_0 =$	**0.1516**	1/min	9.10 (h^{-1})	Eq. V =	−5.60E-09	3.13E-17
		$n =$	**3.62**	–		SS (sum of squares) =		3.29E-10
		$m =$	**3.78**	–				
Equil. mole fraction		$Y_e =$	**0.2028**	–	check $K_L a$	Eq. IV =	0.1476	min^{-1}
Diff mole fraction		$Y_d =$	0.2095	–	exit gas	$Y_{ex} =$	0.1900	–
	Eff. depth	$d_e =$	1.27	m	DO	$C =$	2.00	mg/L
	Depth ratio	$e =$	0.42	–	Check C*inf	$C_\infty^* =$	10.00	mg/L
	Press. at de	$P_e =$	111422	N/m^2				
Input Data		Z_d (m)	Q_a (m^3/ min)	$K_L a$ (1/ min)	C_∞^* (mg/L)	r_w (N/m^3)	H(mg/L/ kPa)	P_{vt} (N/m^2)
		3.00	0.1636	0.1450	10.00	9789	4.425E-04	2333
Equations used								

Eq. 6.7	Eq. I: $K_1 a = (1 - \exp(-mxK_1 a_0 Z_d))/n/m/x/Z_d + (n-1) K_1 a_0/n$
Eq. 6.6	Eq. II: $C_\infty^* = nH \times 0.2095 \times (P_a - P_{vt} - P_d \exp(-mxK_1 a_0 Z_d))/(1 - \exp(-mxK_1 a_0 Z_d))$
Eq. 6.8	Eq. III: $\ln(P_e \ mxK_1 a_0/n/r_w) + \ln(nHY_d P_d/C_\infty^* - 1) = mxK_1 a_0 Z_e$
Eq. 6.4	Eq. IV: $K_1 a = (1 - \exp(-x(1-e) Z_d K_1 a_0))/(x(1-e) Z_d)$
Eq. 6.5	Eq. V: $Y_0 = C_\infty^*/(nH(p_a - p_{vt})) + (0.2095nPd/(n(P_a - P_{vt}) - C_\infty^*/(nH(p_a - p_{vt})) \exp(-HkmZ_d)$
Note: $de = 1/r_w (C_\infty^*/(HYe) - P_a + p_{vt})$	

Next, we consider the case of a mass balance in wastewater where microbes with a respiration rate are present (according to Mahendraker et al. (2005b), the effect of the microbes on the floc is understood to be an increase in resistance to oxygen transfer); in this case, since the resistance is increased, the exit gas Y_{ex} would be increased at $C = 2$ mg/L; hence, $gdp_f = 1.204 \times .1636\,(.2095-.2025) \times 60 = 0.083$ kg/hr/m^2, hypothetically assuming $Y_{ex} = 0.2025$ when the DO value is very close to zero. If the system is at steady state, the gas depletion rate in the air stream must be equal to the respiration rate RV, and, based on Eq. 6.16, the in-process mass transfer coefficient K_La_f is calculated by:

$$K_La_f = (.083 + 0.083)\,(\text{kg/hr/m}^2)/(0.99 \times 10 - 2)/.001\,(\text{kg/m}^3)/60\,(\text{min})/3\,(\text{m}^3/\text{m}^2)$$
$$= 0.1167\ \text{min}^{-1}$$

Therefore,

$$\text{alpha}\,(\alpha) = K_La_f/K_La = .1167/.1450 = 0.80$$
$$R = 0.0830/(3 \times 1) \times 1000 = 27.7\ \text{mg/L/hr}$$

As can be seen, alpha (α) is very sensitive to the exit gas oxygen mole fraction Y_{ex}, so that the off-gas method must be carried out with extreme care in order to obtain a credible alpha value, especially when the exit gas is close to the feed gas mole fraction as can be seen in Hu's (2006) experiment, shown in Fig. 6.3 based on his test results. Small changes in the measurement of the off-gas can give a large error in the estimate of K_La_f. Therefore, from this example, it can be seen that with the concept of a baseline K_La applied to dirty water, a more realistic alpha (α) value can be obtained, bearing in mind that $\alpha = \alpha'$ in the context of this estimation.

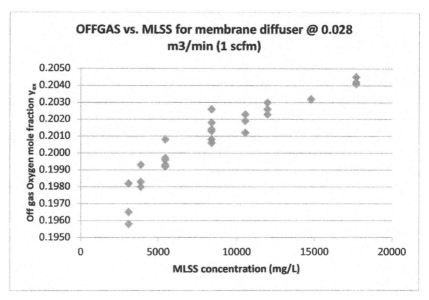

FIGURE 6.3 Off-gas mole fraction vs. MLSS concentrations [Hu 2006]

6.1.5   The Hypothesis of a Microbial Gas Depletion Rate

In the ASCE equation [ASCE 1997] [ASCE 2007], the parameter $K_L a_f$ is used to serve a dual purpose, one to account for the changes in wastewater physical characteristics from clean water to dirty water, but also to account for the variations in the gas depletion rates due to the presence of respiring cells. This equation then makes $K_L a_f$ a variable, dependent on the value of R [Hwang and Stenstrom 1985] because different values of R produces different values of gdp.

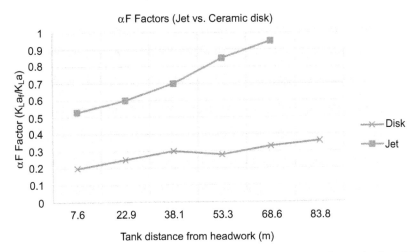

FIGURE 6.4 Comparison of alpha (α) factors for two different microbial respirations [Yunt and Stenstrom 1988c]

As an example, Fig. 6.4 shows the anomaly in the traditional method of determining the ratio of the mass transfer coefficients between in-process water and clean water [Yunt and Stenstrom 1988c].

Although two different aeration equipments and two different wastewater flows were used in the experiments, the ratios should not be so dramatically different, since the clean water tests were carried out comparable in all respects to the process water tests.

Because off-gas measurements in the field tests are reported as OTE_f, it was necessary to translate this value to $K_L a_f$. If the equation used was that of the ASCE 18-96 section 5, given by $K_L a_f = (OTE_f W_{O2} \times 10^3)/(C^*_{\infty f} - C)V$ (where W_{O2} is the mass flow of oxygen in air stream), then this equation has not included the effect of gas depletion which is dependent on R.

The author suspects this difference in αF is more due to the different R values in the field tests. On the other hand, with the new equation (Eq. 6.17), $K_L a_f$ will have only one meaning, which reflects the characteristics of the dirty water only [Eckenfelder 1970] [Stenstrom et al. 1981] [Bewtra and Tewari 1982] [Rosso and Stenstrom 2006a], and is independent of R. Therefore, the wastewater mass transfer coefficient should be given by: $K_L a_f = [(OTE_f W_{O2} \times 10^3) + RV]/(C^*_{\infty f} - C)V$ to conform to the hypothesis that the microbial gdp is the same as the respiration rate.

6.2 MATERIALS AND METHODS USED BY PREVIOUS RESEARCHERS

6.2.1 Garcia et al.'s Experiment [2010]

Using the modified equations and based on the test data by Mahendraker (2003), Garcia et al. (2010) and Hu (2006), it was found that all the testing methods within the ASCE document [ASCE 1997] are valid, as they produce similar values for the mass transfer coefficient $K_L a_f$. In particular, Garcia et al. compared two determination methods for the oxygen uptake rate R, namely the dynamic method and the oxygen profile data method for a fermentation broth. In terms of estimating the OUR and $K_L a$, the methods are similar to the steady-state method in the measurement of OUR, and the non-steady state method in the measurement of $K_L a$, respectively [ASCE 1997]. In the dynamic method example, as described by Garcia and as shown in Fig. 6.5, the airflow inlet to the fermentation broth is interrupted for a few minutes so that a decrease of DO concentration can be observed. When the DO has dropped to an acceptable level, air is turned back on under the same operational conditions until it reaches the same steady state as before. The OUR is determined from the depletion slope after the stopping of the air flow, and the procedure is repeated several times for precision.

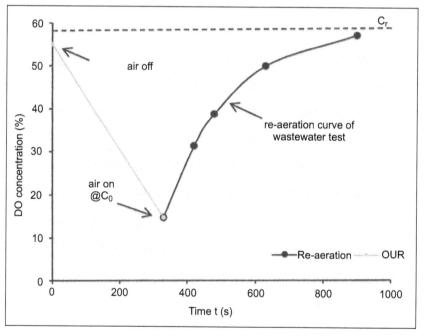

FIGURE 6.5 Dynamic measurement of OUR and $K_L a$ [Garcia et al. 2010]

The second part of the dynamic method is actually identical to the oxygen profile data method that Garcia described, in that both methods require generating an oxygen profile curve. In the dynamic method, the re-aeration is made following the de-aeration by the microbes upon stopping the air supply. (This curve allows the K_La to be calculated). However, the dynamic method requires the OUR to be separately determined, as mentioned in the first part of the test, to be substituted into the oxygenation curve equation to determine K_La. Contrasting with the profile data method where the K_La is pre-determined by other means, including the re-aeration curve, the OUR can be calculated directly from the basic oxygen transfer equation, similar to ASCE Guidelines' equation 2. Therefore, in Garcia's example, since the oxygenation profile is created by re-aerating back to the original DO level, which is similar to the ASCE non-steady state method, which is similar to the profile method, the OUR so determined should be the same as the dynamic method in this example. But it is not (see section 6.2.2 below). The calculation shows that the two R values differ by 50% when comparing the uptake test with the re-aeration test. In addition, Garcia cited experiments by Santos et al. (2006) that showed that in a bio desulphurization microbiological system with the dynamic method employed to measure the OUR, the method differs in the value of OUR dramatically when compared to fitting a metabolic kinetic model to experimental values of oxygen concentration with time. Figure 6.6 and Figure 6.7 compare experimental OUR values obtained from the oxygen concentration profile data (OUR_p) and those obtained using the dynamic method (OUR_d) for two bioprocesses, *Xanthomonas campestris* and *Rhodococcus erythropolis* IGTS8 cultures.

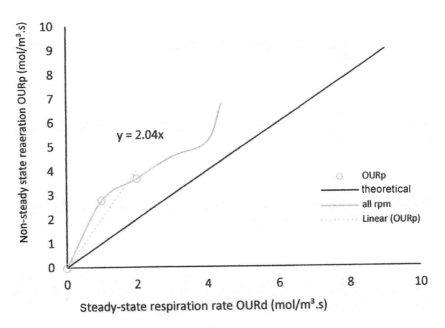

FIGURE 6.6 OUR_p vs. OUR_d (mol O_2/m^3 s $\times 10^{-4}$) [Garcia-Ochoa and Gomez 2009]

As shown in Fig. 6.6, the theoretical relationship between the steady-state test and the non-steady state test should be given by $y = x$ if both tests give the same answer for the mass transfer coefficient, but the actual measurements differ by 50% as seen from the linear relationship $y = 2.04x$. Even though the rotational speed of the magnetic stirrer may differ from test to test, the discrepancy should not be so dramatic. The same holds for the Rhodococcus culture, even though the speeds range from N = 150 rpm to N = 550 rpm. Speed of rotation does affect the K_La and, therefore, affect the prediction of the oxygen uptake rate (OUR), and it can be seen that at higher speeds, the mass transfer coefficient increases dramatically beyond a certain speed. The linear relationship becomes $y = 3.1x$ at 550 rpm, as opposed to $y = 2.3x$ at N = 150 rpm ~ 400 rpm. This effect of speed may be modelled separately, perhaps by adding a scaling factor to K_La pending further experimental investigations. But the effect of gas depletion rate is clear from these experiments, and the effect is additive to the transfer equation as discussed previously, and further below.

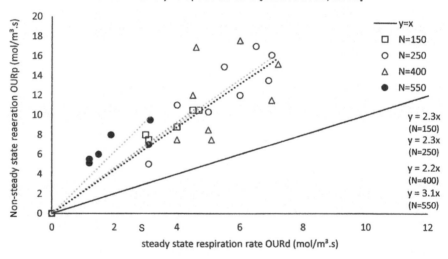

Steady-state vs. non-steady state tests 2006]
Rhodococcus erythropolis culture [Santos et al., 2006]

FIGURE 6.7 OUR_p vs. OUR_d (mol O_2/m^3 s × 10^{-4}) [Garcia-Ochoa and Gomez 2009]

6.2.2   Results and Discussions of Garcia's Experiment

As can be seen from Fig. 6.6 and Fig. 6.7, the experimental OUR values obtained from the DO concentration profile are higher than those by the dynamic method, in fact as much as 100%, depending on the stirring speed (N) of the mixer impeller. The author believes this is due to the mass balance equation in the profile method neglecting the gas side oxygen depletion, thus giving an erroneous OUR value that is twice the value obtained in an *in-situ* oxygen uptake experiment for the same K_La. When the OUR = R, based on the premise that the $OTR_f = K_La_f.(C^*_{\infty f} - C) - R$, this concept leads to modification of ASCE equation 2 for a batch test.

Therefore, to illustrate the anomaly, using Garcia's equation for the re-aeration in Fig. 6.5,

$$dC/dt = K_La \cdot (C^* - C) - OUR$$
$$C = Cr - (Cr - C_0) \exp(-K_La \cdot t)$$

From Fig. 6.5, the following can be read:

$$Cr = 59.5\%; \ C_0 = 14.7\%, \text{ therefore,}$$
$$C \approx 59.5 - (59.5 - 14.7) \exp(-0.0052 \cdot t)$$

where K_La is found to be 0.0052 s^{-1} by fitting the read data to the model by the non-linear least square (NLLS) method, using the Excel solver as shown in the Table 6.2 below:

TABLE 6.2 Reproduced data from Garcia-Ochoa and Gomez [2009]

Start Time	Time from Start (s)	Duration (s)	c (%)	Cr, K_La, C_0 Paramtrs	c(model fit)	Error (c-c(m))	SS
330	330	0	15	59.49205	14.68	0.322	0.103
	420	90	30	0.005168	31.35	−1.347	1.815
	480	150	40	14.67843	38.85	1.149	1.319
	630	300	50		49.99	0.015	0.000
	900	570	57		57.14	−0.137	0.019
						min. sum(SS)	3.257

Garcia et al.'s own calculation in their report gave: $K_La_f = 0.0057$ s^{-1}

If the steady state concentration is taken to be around 55%, then R would be calculated as:

$$R = 0.0057 \cdot (100 - 55)/100 = 0.256\% \ s^{-1}$$

From the oxygen uptake rate test,

$$dC/dt = -R$$

$R \approx (55 - 13)/335 \times 100 = 0.125\%$ s^{-1} which is only half the value calculated by the profile method. If this measured uptake rate is inserted in the above equation as a known quantity, the value of K_La_f obtained is only one-half of that obtained by the NLLS method (the profile method), and will not be correct, since Garcia's formula did not include the effect of gas depletion.

This concept of accounting for gas depletion, based on the depth correction model Eq. 6.17 above, is at first seemingly counter-intuitive, as one of previous reviewers had mentioned: "In the text is indicated that "the higher the reaction rate, the smaller the gas depletion rate". This phrase is difficult to understand because in an aerobic process if reaction rate increased, the oxygen consumption will be higher (the oxygen is needed to degrade organic matter) and gas depletion should be *higher"*, but can be readily understood when a gas phase mass balance of oxygen is taken for a liquid volume when the system is at steady state. The difference between the feed gas rate and the exit gas rate must be the oxygen transfer rate (OTR), which is equal to R, but the OTR is also the gas depletion rate, and so the microbial *gdp* must also be equal to R.

The text simply means that for *the same gas supply rate* (therefore constant $K_L a_f$ during the duration of the study), an increase of R, such as an organic shock load, adds an additional resistance and so the microbial *gdp* would *increase*, but the overall *gdp* or OTR would *decrease* (see Fig. 6.3), requiring the system to adjust to a new steady-state by lowering the steady state DO concentration C_R, thereby increasing the driving force so that the OTR_f would match the new oxygen demand. However, if C_R becomes too low, the blowers might then need to work doubly hard, not only to constantly provide enough air to maintain the oxygen being consumed by the biomass (oxygen uptake rate $OUR_f = R$), but also to maintain a stable 'spare' DO level required to overcome the additional resistance. In this case, the *gdp* would obviously increase to counteract the increase in R, but the gas flow rate Q_a would also be different and then it would violate the limitations of the test [ASCE 1997], as change in the gas flow rate means that $K_L a_f$ is no longer constant. In the experiment on the performance of a membrane bioreactor (MBR) treating high strength municipal wastewater, conducted by Birima et al. (2009), the results of dissolved oxygen (DO) and aeration rate show that the effect of the organic loading rate (OLR) on aeration rate and DO concentration was very significant. For instance, comparing the results of a trial with low OLR with those of another trial with high OLR shows that the aeration rate in the first trial was 20 L/min corresponding to DO of above 4 mg/L, whereas the rate of aeration in the second trial increased rapidly till 60 L/min but corresponding to a DO of below 2 mg/L. Similarly for other trials, it was noted that the higher the organic loading rate, the higher would be the aeration rate and correspondingly the lower the DO concentration. This observation appears to support the hypothesis of a higher resistance to oxygen transfer when the demand for oxygen has increased, even though the driving force has increased because of the lowered DO. This implies that for higher organic load, a higher rate of aeration is required to obtain the same DO. This means that operating the MBR with a high organic load means that more energy is required. Generally, the results of their study showed that for the low OLR trials, the aeration rate varied from 6 to 12 m^3/m^2 membrane area per hour and the DO varied from 3.7 to 5.7 mg/L, whereas for the high OLR trials the aeration rate and the DO varied from 6 to 18 m^3/m^2 membrane area per hour and 0.9 to 4.4 mg/L, respectively. This depends on the concentration of MLSS in the reactor that in turn directly affects the respiration rate of the microbial communities.

6.2.3   Other Previous Test Results Re-visited and Discussions

6.2.3.1   *Yunt et al.'s (1988a) Reported Data*

Experimental verification to justify the depth correction model for a batch mode in clean water is given by Lee [2018] and Yunt et al. (1988a) experiments have been described in Chapter 3 Section 3.3. The test results are given in the LACSD report Table 5: "Summary of Exponential Method Results: FMC Fine Bubble Tube Diffusers" and copied over as shown by Table 5.8 in Chapter 5 as well as Table 3.1 in Chapter 3, where the calculations for the baseline $K_L a_0$ are shown by Table 3.2. The calculation spreadsheet for estimating the variables $K_L a_0$, n, m, de and ye is not repeated in this chapter. Using the standardized baseline, $(K_L a_0)_{20}$, the simulated result for a typical run test for the FMC diffusers gives a value of $(K_L a)_{20} = 0.1874$ min^{-1} as compared to the test-reported value of $(K_L a)_{20} = 0.1853$ min^{-1} which gives an error difference of around 1% only compared to the simulated value [Lee 2018]. The compared results of the aeration efficiencies plotted in ascending order of the tank depths is shown herewith in Fig. 6.8 (which is identical to Fig. 3.9 in Chapter 3). Within experimental errors and simulation errors, the results seem to match very well, with the aeration efficiency slightly over-predicted at the higher depths. (This is probably because in the development of the model, any free water surface oxygen transfer has been ignored. This effect is not uniform for all tanks – it is more important for shallower tanks than for the deeper ones [DeMoyer et al. 2002], and so the actual baseline would have been slightly smaller for the deeper tanks if the effect of surfacing bubble plume had been considered, matching the report values). It would appear from the graph that the oxygen transfer efficiency is an increasing function of depth, even though the gas flow rates were not exactly the same for all the tests. These results, along with other tests, clearly show that the Lee-Baillod model [Lee 2018] is valid in considering the effect of oxygen gas depletion for the purpose of predicting oxygen transfer.

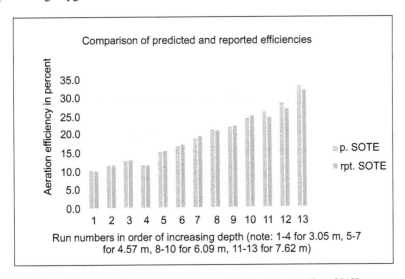

FIGURE 6.8 Simulated efficiencies for FMC diffusers [Lee 2018]

As shown in Fig. 6.2 for the depth correction model given in Section 6.1.3, $K_L a_f$ is a declining trend with respect to increasing depth of the immersion vehicle of gas supply, similar to what Boon (1979) has found in his experiments. The detailed analysis and the derivation of the model equation (Eq. 6.17) is given by Lee [2018] and also in Chapter 4. For similar tests in in-process water, the mass transfer coefficient so determined by the depth correction model while incorporating the alpha factor should match the result of field testing by following the ASCE Standard Guidelines for In-Process Oxygen Transfer Testing methods [ASCE 1997]. The verification of validity for this equation would require testing full-scale in the field, using the alpha' value (α') derived by bench-scale testing or shop tests, with solids filtered out or the cells destroyed prior to testing.

6.2.3.2   Mahendraker (2003) Data

Based on re-analyzing the data from Mahendraker's dissertation [Mahendraker 2003], the average value of α' based on all the test results is about 0.82. The new equations after incorporating the effect of *gdp* give an overall discrepancy for $K_L a_f$ of only around 12% between the various tests, notwithstanding the various inaccuracies of the individual methods stated in the literature [ASCE 1997]. The calculations are not shown in this paper, but readers can satisfy themselves by verifying the results and by examining their paper and re-analyzing the experimental data. It should be noted that, in calculating for $K_L a_f$ in their steady-state tests, the new equation Eq. 6.16, as postulated by the author and as repeated below, was used instead of the conventional model:

$$K_L a_f = \frac{2R}{(C^*_{\infty f} - C_R)}$$

In this equation, although R was determined by the respirometer method as described in Mahendraker's dissertation, the author considers this method as incorrect in that the dissolved oxygen content in the sample was artificially aerated to a higher level, thus in the author's opinion artificially making the estimation of the microbial respiration rate twice as much as would be in the actual treatment system. R must therefore be reduced by 50% in the mass balancing equations for calculating the mass transfer coefficients in these tests.

Furthermore, Mahendraker et al. (2005b) postulated that the resistance to oxygen transfer is composed of two elements: the resistance due to the reactor's solution, and the resistance due to the biological floc. They formulated the relationship between these resistances as:

$$\frac{1}{\propto K_L a} = \frac{1}{\propto' K_L a} + \frac{1}{K_L a_{bf}} \tag{6.20}$$

in which the scaling factor for the reactor solution was given in their equation as $\propto_e$, instead of the symbol α' and the subscript *bf* denotes biological floc. Hence, the resistance due to the biological floc is given as $1/K_L a_{bf}$. Rearranging Eq. 6.20 gives:

$$\propto K_L a = \frac{\propto' K_L a \times K_L a_{bf}}{\propto' K_L a + K_L a_{bf}} \tag{6.21}$$

Substituting Eq. 6.21 into Eq. 6.2 which is identical to ASCE 18-96 equation 2, in a steady state and for a batch reactor, and solving for $K_L a_{bf}$ we have:

$$K_L a_{bf} = \frac{\propto' K_L a \times R}{\propto' K_L a (C^*_{\infty f} - C_R) - R} \tag{6.22}$$

If it is assumed that the resistance of the biological floc is the same as the resistance from the reactor solution (this is similar to assuming that the gdp due to the microbes is due to the resistance in the bioreactor solution $K_L a_f$), then

$$K_L a_{bf} = \propto' K_L a \tag{6.23}$$

Substituting Eq. 6.23 into Eq. 6.22 and solving for $\propto' K_L a$,

$$\propto' K_L a = \frac{2R}{(C^*_{\infty f} - C_R)} \tag{6.24}$$

This equation is similar to Eq. 6.16 previously derived. However, this requires an assumption for the biological floc resistance which may not be true, as well as the use of the ASCE equation that the author has disputed its validity. Mahendraker et al. (2005b) concept requires further investigation. The logical explanation may be that the biological floc resistance results in a change of the gas depletion rate, which coincides with the microbial respiration rate at steady state. This explanation then unifies the two concepts together, and result in the same conclusion as stated by Eq. 6.24.

6.3 RELATIONSHIP BETWEEN ALPHA (α) AND APHA' (α')

As mentioned in Section 6.1.3, the concept of a baseline mass transfer coefficient for wastewater requires the employment of an additional correction factor (α') for the clean water $K_L a$ when applying the standard model to wastewater. However, alpha (α) and alpha' (α') are inter-convertible. From Eq. 6.13, we have,

$$\frac{dC}{dt} = K_L a_f (C^*_{\infty f} - C) - gdp_f - R$$

Therefore, substituting $\alpha' K_L a$ for $K_L a_f$, we have,

$$\frac{dC}{dt} = \alpha' K_L a (C^*_{\infty f} - C) - gdp_f - R \tag{6.25}$$

Therefore, if the microbial gas depletion rate is the same as the respiration rate, we have

$$OTR_f = K_L a \cdot \alpha' \,(\text{deficit}) - R \tag{6.26}$$

where deficit = $C^*_{\infty f} - C$

For the case where alpha is directly determined in the field by comparing in-process wastewater to clean water tests simultaneously, as described in the previous sections,

$$OTR_f = K_L a \cdot \alpha \,(\text{deficit}) \tag{6.27}$$

Equating Eq. (6.26) and Eq. (6.27), alpha and alpha' can be inter-related as:

$$\alpha' = \alpha + \frac{R}{K_L a (\text{deficit})} \tag{6.28}$$

Both Eq. (6.26) and Eq. (6.27) can be used to determine the oxygen transfer rate under process conditions. However, Eq. (6.27) has two degrees of freedom, with both variables (water characteristics and cell respiration) incorporated into this one single value for alpha. Alpha values can vary from a small value to a large number, depending on the initial cell content and the degree of treatment which directly affects the value of R. It is also dependent on the organic loading rate (oxygen demand) to which the metabolic rate (consumption) directly responds.

Alpha', on the other hand, depends only on the nature of the wastewater, which can be much more constant. The two equations are easily reconciled by substituting Eq. 6.28 into Eq. 6.25, giving

$$\frac{dC}{dt} = \left(\alpha + \frac{R}{K_L a(\text{deficit})} \right) K_L a(\text{deficit}) - gdp_f - R \tag{6.29}$$

or

$$\frac{dC}{dt} = \alpha K_L a (C^*_{\infty f} - C) - gdp_f \tag{6.30}$$

Again, if gdp is equal to R, we have

$$\frac{dC}{dt} = \alpha K_L a (C^*_{\infty f} - C) - R \tag{6.31}$$

which is identical to ASCE equation 2 in the guidelines [ASCE 1997], for a batch process.

In the application of Eq. (6.26) or Eq. (6.27), both equations should give similar results if the objective is to find the oxygen transfer rate under process conditions, given clean water test results for $K_L a$ and C^*_∞ (note that C^*_∞ is required to determine the deficit). Eq. (6.26) would additionally require the determination of alpha', as well as determination of R which can be done in accordance with the ASCE 18-96 methods. On the other hand, Eq. (6.27) does not contain an R term; therefore, Eq. (6.27) is the more popular choice when the objective is to determine the OTR_f directly because of its simplicity.

However, if the objective is to find R, the oxygen uptake rate commonly known as the OUR, Eq. (6.27) cannot be used even though all the other parameters are known, since alpha has two degrees of freedom, and so it is impossible to determine how much of the transfer is attributed to R and how much is due to the gas-liquid transfer mechanism that is affected by the water characteristics. Additionally, the intensive variable of the difference in gas depletion rates is also unknown. With three unknowns, R cannot be calculated even if all the other parametric values are known in (Eq. 6.27). R must therefore be independently measured.

6.4   MEASUREMENT OF RESPIRATION RATE

According to ASCE guidelines (ASCE 1997, 2018), measuring the OUR under oxygen limiting conditions is extremely difficult, although the principle of respirometry is remarkably simple—-the slope of the decline curve of DO vs. time must be the respiration rate. The problem is not so much the method as the methodology commonly employed to make the sample measurable. When the oxygen level in the sample is low, say, at 2 mg/L, it would need to be artificially aerated to a higher level, say, 5 mg/L, before a meaningful curve (usually a straight line if the sample is not substrate-limiting as well) for calculating the slope can be obtained. This boosting of the DO concentration may make the sample measurement artificially high, and so the true uptake rate in the aeration tank is not measured correctly. How do we correct this error?

To estimate the respiration rate R, ASCE (2018) recommends the off-gas column steady-state test. In the recommendation, an acrylic or fiberglass reinforced tank is used, such as a 30 in. (760 mm) diameter by 11 ft (3.4 m) deep column. The column depth was selected based on work done at the University of Wisconsin (Doyle 1981) where it was found that alpha decreased as the liquid depth increased over a range of two to ten feet (3.05 m); however, the decrease was relatively small above eight feet (2.4 m). Mixed liquor is continuously pumped to the test column from a position within the existing aeration tank using a submersible pump. The liquid detention time in the column is typically maintained between 10 and 15 min. The mixed liquor should be aerated using a fine pore (fine bubble) diffuser identical to the type used in the tank. The oxygen transfer efficiency of the diffuser used in the column using process mixed liquor is measured using the off-gas techniques described in Section 3.0 of the Guideline. The airflow rate to the test diffuser is adjusted so that the DO concentration in the steady state column is maintained in the range of those found in the test section of the aeration basin. A schematic of the column test system is given in Figure D-1 of the ASCE Guidelines.

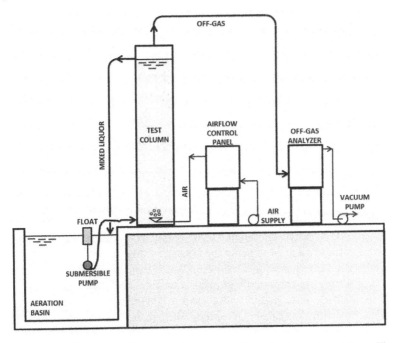

FIGURE 6.9 Mixed liquor steady state column test schematic (reproduced from Fig. D.1 ASCE 18-96)

An example is given in the Guidelines as shown below:

An *ex situ* column test is performed at a test section of the aeration basin. The following data are collected:

DO_i (at transfer pump) = 0.55 mg/L

DO_o (in test column) = 0.80 mg/L

qi (airflow rate to column) = 1.07 NL/s (normal liters per second)

OTE_f (measured in column) = 0.130 mg O_2 transferred/mg O_2 supplied

Q_i (mixed liquor pump rate) = 2.16 L/s

V (column volume) = 1,460 L

OTR_f = 1.07 NL air/s × 299.3 mg O_2/ NL air × 0.13 mg O_2 transferred/

$$mg\ O_2\ supplied = 41.6\ mg\ O_2/s$$

Oxygen uptake rate is determined by a mass balance of oxygen around the column system as:

oxygen uptake rate = (oxygen transfer rate – net change in DO)/column volume,

or $R = (OTR_f - (DO_o - DO_i)Q_i)/V$

R = [41.6 – (0.80 – 0.55) × 2.16]/1,460 = 0.0278 mg/L·s × 3600 s/hr = 101 mg/L/hr

Although, unfortunately, there is no comparable data using the other methods such as the BOD bottle method, Chiesa et al. [1990] conducted a series of bench- and pilot- scale experiments to evaluate the ability of biochemical oxygen demand (BOD) bottle- based oxygen uptake rate (OUR) analyses to represent accurately *in-situ* OUR in complete mix- activated sludge systems. Aeration basin off- gas analyses indicated that, depending on system operating conditions, BOD bottle-based analyses could either underestimate *in-situ* OUR rates by as much as 58% or overestimate *in-situ* rates by up to 285%. A continuous flow respirometer system was used to verify the off- gas analysis observations and assessed better the rate of change in OUR after mixed liquor samples were suddenly isolated from their normally continuous source of feed. OUR rates for sludge samples maintained in the completely mixed bench- scale respirometer decreased by as much as 42% in less than two minutes after feeding was stopped. Based on these results, BOD bottle- based OUR results should not be used in any complete mix- activated sludge process operational control strategy, process mass balance, or system evaluation procedure requiring absolute accuracy of OUR values. This echoes the author's previous suspicion about Eckenfelder's experiment (Eckenfelder 1952) for calculating the oxygen transfer efficiency based on sample testing of the respiration rate which required artificial boosting of the DO concentration in a BOD-bottle, as described below.

There is certainly a need to get to the bottom of this. According to the published article, "Aeration Efficiency and Design", in which Eckenfelder (1952) described two methods of testing for the microbial respiration rate, both the steady-state method and the non-steady state method were used to "validate" that these two methods are compatible with each other. The results were given in a table as reproduced and summarized below (Table 6.3).

TABLE 6.3 Laboratory test results by Eckenfelder (1952)

Run No.	AFR (Air Flowrate)			NSS	log-def	SS	rpt
	Q(cfh)	Q (m³/min)	Q_a (m³/min)	calc. $K_L a$ 1/h	rpt $K_L a$ 1/h	$K_L a$ (ASCE) 1/h	R (mg/l/h)
	0					0	
C1	33	0.0156	0.0147	8.03	8.20	8.50	27.5
C2	39	0.0184	0.0173	11.31	9.80	9.37	30.5
C3	52	0.0245	0.0231	12.21	12.00	11.96	31.5
C4	64	0.0302	0.0284	16.01	15.00	15.89	34.4
C5	77	0.0363	0.0342	18.99	18.20	18.78	33.4
C6	92	0.0434	0.0409	24.87	23.50	22.70	32.2

In this table wereshown the test results for 6 runs, where cfh stands for cubic feet per hour for the air supply, for an aeration tank of 33 inches (838 mm) tall, and aerated at different flow rates from 33 cfh (0.016 m³/min) to 92 cfh (0.043 m³/min). In terms of SI units, the air flow rate (AFR) and the height-averaged air flow rate Q_a would essentially be the same given the small height. Eckenfelder used the log-deficit method to calculate the mass transfer coefficients, corresponding to each AFR, for the non-steady state (NSS) test results, shown in red in col. 6. His data was reproduced and converted and then the non-linear regression analysis (NLLS) was used as recommended in the standard (ASCE 2-06) to re-calculate the $K_L a$'s, shown in col. 5. The plot of the re-aeration curves is as shown in Fig. 6.10. These

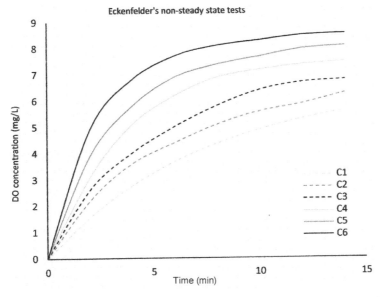

FIGURE 6.10 Illustrative problem in oxygen transfer measurement, reproduced from Eckenfelder (1952)

data were derived from an experimental aeration tank employing an agitator and air sparger ring. Next, the author used the steady-state (SS) method to again calculate K_La, after using the measurements of the individual respiration rates from the BOD bottle method, and equating those with the oxygen transfer rate as $K_La \, (C^* - C)$, and the results, as shown in col. 7, are similar to those of the non-steady state method, ostensibly proving the validity of both test methods. Unfortunately, clean water tests were not performed, and so there is no way to estimate alpha. However, the respiration tests were done at 2 p.p.m. as reported in the article by Eckenfelder, and so these samples must have been re-aerated by vigorous shaking to at least twice the value of the *in-situ* dissolved oxygen concentration.

If the author's hypothesis is correct, i.e. that the microbial oxygen uptake rate is linearly proportional to the oxygen availability that reflects the amount of agitation, then the measured R values must have been at least twice the actual values in the aeration basin where the samples were withdrawn. The resultant K_La values would then be half the actual values measured by the non-steady state method. This experiment then in fact did not prove the validity of either one or the other, but instead proved that these two methods give results of K_La that are off by 50% using the ASCE methods.

According to the author's thesis, the equation for the steady-state method should come out to be $K_La_f = 2R/(C^* - Cr)$ instead of a single R as conventionally used because of the gas depletion effect in the air bubbles. With this modification, this would then give the exact results of the K_La_f as before using the ASCE non-steady state method, if it is reckoned that the BOD method of measuring the OUR is incorrect (over-estimation) because of the additional aeration. It is therefore vital, to prove one way or another, that an *in-situ* oxygen uptake rate test be performed similar to that described in the Guideline ASCE 18-96 (or more recently ASCE 18-18) for the steady-state column test using the off-gas measurement techniques. This may confirm, once and for all, whether oxygen availability has an effect on the oxygen uptake rate in a sample upon re-aeration.

The beauty of this off-gas method, as shown by Fig. 6.9, in measuring the respiration rate is that it does not require artificially aerating the sample to a higher DO level, and K_La doesn't come into the picture as well. The author suggests that an experiment be done in one treatment plant with an acrylic column 3 m or so high, and then comparing the result with the traditional BOD bottle method and observing the difference, especially for oxygen-limiting low DO conditions. The key element of success is the off-gas analyzer that must measure the off-gas accurately, since the OTE is highly sensitive to changes in the gas composition (see Chapter 5 above).

Measuring OUR under oxygen limiting conditions is difficult, and it was found that OUR measurements in general can be impacted by how the test isrun [Doyle 1981] [Private communication]. When one withdraws a sample of mixed liquor and runs an OUR the typical way (aerate to high DO in a BOD bottle, stop aerating and then track the depletion of DO over time), one changes the conditions in the sample compared to the conditions in the aeration tank. Of course, that method is not at all appropriate for an aeration basin near zero DO (i.e. the tank OUR is limited by the oxygen transfer rate). When more DO is given, the bacteria will increase the OUR compared to the oxygen limited OUR in the aeration tank. But even with sufficient aeration basin DO, one can get different OUR values depending on the measurement method. For example, shaking the sample to aerate can break up the floc, making the substrate and DO more available to the bacteria. Doyle noticed in one of his thesis work that a BOD bottle OUR did not agree with a respirometer OUR. When using the old Arthur respirometer, it consistently gave higher OUR measurements compared to the BOD bottle/DO depletion method. He attributed it to the intensive aeration in the respirometer which was rather violent and may have broken up the floc causing increased delivery of DO and substrate to the floc [Private communication].

In Garcia-Ochoa et al. (2010) experiment (see section 6.2.2), Garcia tried to explain this anomaly by the "cellular economy principle" that, during the time oxygen is not transferred (i.e. during the shutting off the gas supply in the microbial desorption test), microbial cells consume oxygen at a lower rate. There are four reasons that this claim is wrong:

  i. the desorption test is done *in-situ*, there is no time lag between DO (dissolved oxygen) analysis and sample collection;

 ii. The desorption curve is linear which means the decrease in DO content is uniform (see Figure 6.5). This in turn means the microbes are consuming the oxygen at a uniform rate. If the microbes had been using oxygen at a declining rate, the curve would have been concave in shape;

iii. During desorption, there is still plenty of oxygen in the liquid phase, beginning at 55% saturation. Bacteria are not so smart that they could sense a continual diminishing of oxygen that they would start economizing from the start. Only when the DO has reached the endogenous zone would this occur. Even if the consumption rate has decreased, it could never by as much as 50%.

 iv. There is no deficiency in soluble substrate and so the respiration rate would not be affected by soluble substrate uptake depletion, i.e. the system was not at substrate-limiting condition before or during the test.

According to Doyle, there is no data to show that isn't the case that the microbes change their respiration rate at low DO levels, but it seems to Doyle that oxygen is used as fast as it can be delivered. Any reduction is due to physical/chemical limitations (diffusion limitations, and concentration gradients and transport across the cell membrane). It would be difficult to determine whether the microbes are doing this voluntarily since to test the theory one has to limit the DO, which impacts the other factors, such as an increase in the biological floc respiration rate due to the breaking up of the floc by violent agitation in order to bring the DO level back up to a measurable level for testing.

In summary, the author reckons there are only three possible explanations for an over-estimation of the respiration rate. Firstly, when the sample is agitated by vigorous shaking, the activity level of the microbes might increase. Like any living organismsunder stress, they respire more. Secondly, the oxygen level in the sample may be limiting (although at 2 ppm, it shouldn't be); thirdly, Garcia-Ochoa et al. (2010) suggested a cell economy principle by which the microbes voluntarily reduce the respiration level at low DO and changes the respiration rate at elevated level due to increased oxygen availability. There have been many literature on this but none seems to have given a definite answer.

However, Doyle's article [Doyle 1981] suggested an interesting method of testing for alpha. It seems that it may be possible to use a dilution method to test out the determination. By first aerating a tank of pure water to an elevated DO, say 7 ppm, then pouring the activated sludge mixed liquor into the tank, and then gently mixing them together, it may be possible to measure the slope of the DO decline curve at quiescent conditions, thereby eliminating the first possible explanation for the cause of increased OUR measurement. If the sample has been diluted to 50%, the resultant slope should then be multiplied by 2 to get the true OUR. This should then be compared with a corresponding steady-state column test with an *insitu* measurement.

6.5   CONCLUSIONS

In this manuscript, the author postulates that the true OTR_f is given by $K_L a_f (C_f^* - C)$ $- R$, since the transfer rate is affected by biochemical reactions in the cells [Hwang and Stenstrom 1985], which changes not only the water characteristics but also the *gdp*. Based on the studies so far cited in this manuscript, it is concluded that:

1. The oxygen transfer efficiency based on the oxygen transfer rate by a prescribed CWT (Clean Water Test) for a fixed gas supply is a property of an aeration equipment, and sowill not be affected by external factors (i.e. clean water test data are reproducible) [ASCE 2007] and would be uniquely defined by a standard specific baseline value $K_L a_0$;

2. A new mathematical model for *gdp* has been derived (Eq. 6.17) and is verified by testing under a variety of water depths for clean water (Yunt et al. 1988a). This model is shown to be applied to wastewater through a revised alpha (α') that pertains specifically to wastewater characteristics;

3. The respiration rate produces additional resistance to oxygen transfer in the system resulting in a loss of *gdp*, and therefore must be accounted for in the mass balancing equations. In this sense, the OTR_f in the system (as opposed to the aeration efficiency of the device definable by the CWT) is indeed affected negatively by the OUR;

4. The difference in the gas depletion rates due to the microbial cells that affect oxygen transfer is precisely the respiration rate itself, based on all the test results, and therefore is important for the revising of the ASCE equations [Mahendraker 2003, Mahendraker et al. 2005a];

5. The present equations used in the ASCE Guidelines [1997] are not correct for submerged aeration. This has resulted in discrepancies of around 40 ~ 50% in the estimation of $K_L a_f$ for batch test analyses (i.e. the steady-state test results are lower than the non-steady state tests by such). The new equations after incorporating the effect of *gdp* give an overall discrepancy of only around 12%, notwithstanding the various inaccuracies of the individual methods stated in the literature [ASCE 1997] [Mahendraker 2003] and for the continuous water flow testing;

6. K_La_f is dependent on the AFR (air flow rate), but since the AFR also affects the corresponding K_La in clean waters, the resultant effect on α is not overly significant when α is defined as α'. The average value of α' based on all the test results is about 0.82 according to Mahendraker's data. This value of α' is in line with the traditional design value of 0.8 used in many treatment plant's designs. However, using this value would now require revising the design equations to include the gas depletion effect as explained in this manuscript; otherwise, α would become highly variable [Rosso and Stenstrom 2006a, b] [Jiang et al. 2017].

As a consequence of including the gas depletion effect in the application of Clean Water Test Results to estimate Oxygen Transfer Rates in Process Water at Process DO levels, for the same oxygenation system, the following amendment is applicable to Eq. CG-1 in ASCE 2-06 [ASCE 2007]:

$$OTR_f = \left(\frac{1}{(C_\infty^*)_{20}}\right)[\alpha(SOTR)\theta^{T-20}](\tau\cdot\beta\cdot\Omega\cdot(C_\infty^*)_{20} - C) - RV \qquad (6.32)$$

where $\alpha = \alpha$ in the context of this submitted document, with all symbols referring to the ASCE (2007) Standard. The implication of this equation is that the oxygen transfer in the field of an aeration equipment can be closely determined by clean water tests by applying relevant correction factors to the clean water measured parameters, together with accurate measurements of the respiration rate in the field. To complete the equation, the effects of temperature in the selection of a proper value for the temperature correction parameter θ, and the effect on K_La due to geometry have been discussed in previous chapters and manuscripts [Lee 2017, 2018].

6.6   APPENDIX

6.6.1   The Lee-Baillod Model in Wastewater (Speculative)

If the clean water K_La is to be applied to in-process water, this discrepancy in the gas depletion rate between the two systems must be accounted for in any mathematical model describing oxygen transfer in in-process water, so that the meaning of the mass transfer coefficient is consistent with both systems. It is thought that the mathematical model, starting with clean water at the equilibrium state where the parameters (K_La and C_∞^*) describe the oxygenation curve that can be determined in a clean water test, might be applied to wastewater. The derivation of such a model for clean water has been published in the WERF Journal [Lee 2018], and also explained in detail in Chapter 4. It is envisaged that the standard specific baseline for wastewater can be similarly determined as for clean water, pending further investigations.

6.6.2   Determination of the Standard Specific Baseline for Wastewater (Speculative)

Revisiting Yunt et al.'s experiment [Yunt et al. 1988a, 1988b] in a shop test, Fig. 6.11 shows that the resulting K_La_0 values [Lee 2018] obtained for various tests are adjusted to the standard temperature by the temperature correction equation of the 5th power model (Lee 2017) and plotted against Q_{a20}.

This curve is identical to Fig. 5.7 in the previous chapter, except that the unit for the baseline mass transfer coefficient is in 1/hr where previously it was in 1/min. Remarkably, all curves fitted together after normalizing K_La_0 values to 20°C, as shown by the data points (represented by different symbols) on the different depths. The exponent determined is 0.82. The value obtained from the slope is 44.35 × 10^{-3} (1/min) for all the gas rates normalized to give the best NLLS (Non-Linear Least Squares) fit, bearing in mind that the K_La_0 is assumed to be related to the gas flowrate by a power curve with an exponent value [Rosso and Stenstrom 2006a] [Zhou et al. 2012]. The slope of the curve is defined as the standard specific baseline. Therefore, the standard specific baseline (sp. $K_La_0)_{20}$ is calculated by the ratio of $(K_La_0)_{20}$ to Q_{a20} 82 or by the slope of the curve in Fig. 6.11.

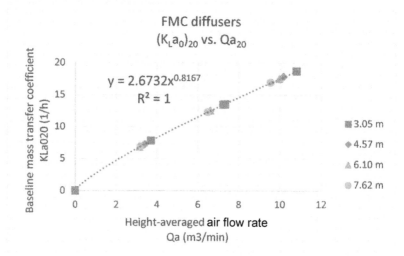

FIGURE 6.11 Effect of diffuser submergence and airflow rate on the baseline transfer coefficient

When the same information is compared with a similar plot using the actual measured K_La values (plot shown as Fig. 6.12), it can be seen that the correlation was still quite good for the curve, but not as exactly as when the baseline values were plotted, testifying to the fact that the baseline mass transfer coefficient does represent a standardized performance of the aeration system when the tank is of zero depth (i.e. when the effect of gas depletion in the fine bubble stream was eliminated). As K_La is a local variable dependent on the bubble's location, especially its height position, K_La_0 represents the K_La at the water surface, i.e. at the top of the tank with no gas depletion, where the saturation concentration corresponds to the atmospheric pressure ($P_S = 1$ atm).

By the same token, it is speculated that the same correlation would exist for wastewater, so that a **standard specific field baseline** $(K_La_{0f})_{20}$ **per** Q_{a20}^q can be equally established by similar testing on wastewater, *provided that any gas depletion effect from microbial respiration is avoided.* The wastewater mass transfer coefficient $K_L a_f$ can then be calculated by Eq. 6.17, after the baseline $K_L a_{0f}$ has been determined by such testing or by laboratory bench-scale testing.

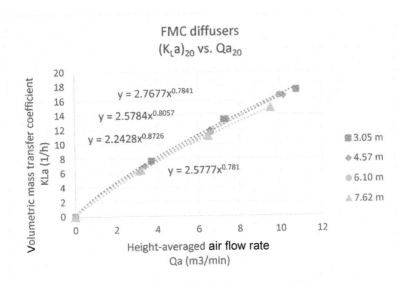

FMC diffusers
$(K_L a)_{20}$ vs. Qa_{20}

$y = 2.7677x^{0.7841}$

$y = 2.5784x^{0.8057}$

$y = 2.2428x^{0.8726}$

$y = 2.5777x^{0.781}$

- 3.05 m
- 4.57 m
- 6.10 m
- 7.62 m

Volumetric mass transfer coefficient
KLa (1/h)

Height-averaged **air flow rate**
Qa (m3/min)

FIGURE 6.12 Effect of diffuser submergence and airflow rate on the mass transfer coefficient

If the Lee-Baillod model (Eq. 6.5) is applied, then the water characteristics, such as E (modulus of elasticity), ρ (density of the wastewater), and σ (surface tension of the wastewater) must be separately determined in order for the model to be applicable in the temperature correction model (Eq. 2.1). Also, the Henry's Law constant for the wastewater would be different from that of clean water, so that the parameter x given by $x = HRST/U_g$ needs to be adjusted accordingly (x has been defined as the gas flow constant in Chapter 4). The solutions for $K_L a_f$ by the simulation model should match actual field-testing results using the methods described in the ASCE Guidelines, provided the OTR_f and $K_L a_f$ are calculated by Eq. 6.32 that would include the respiration rate separately determined in the field. The concept of additional resistance from the microbes can be further illustrated by the following mathematical derivation.

6.6.3 Mathematical Derivation for In-process Gas Transfer Model

The various cases are examined as below:

(i) Baseline (R = 0)

First consider the baseline case. For the simple case where oxygen uptake rate is zero, ASCE (Eq. 2.2) or Eq. 6.2 based on a mass balance on the liquid phase gives:

$$\frac{dC}{dt} = K_L a_f (C_{\infty f}^* - C)$$ (6.33)

Based on a mass balance on the gas phase gives:

$$F = K_L a_f (C_{\infty}^* - C)$$ (6.34)

where F is the gas depletion rate per unit volume given by Figure 6.13, where,

$$F . V = \rho_i q_i Y_i - \rho_e q_e Y_e$$

where ρ is density of the gas; q is the gas flow rate; subscripts i and e are inlet and exit.

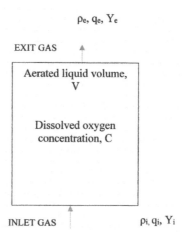

FIGURE 6.13 Mass balance on the gas phase

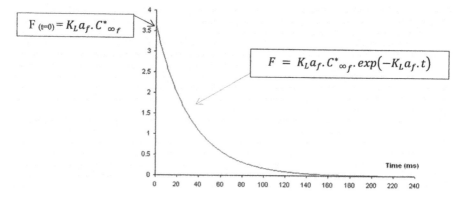

FIGURE 6.14 Gas depletion time variation ($R = 0$)

Simplifying the case by assuming the test starts at zero DO, and integrating Eq. (6.33) gives:

$$C = C_{\infty_f}^* (1 - \exp(-K_L a_f \cdot t)) \qquad (6.35)$$

Substituting C from the above expression into Eq. (6.33) gives:

$$\frac{dC}{dt} = K_L a_f (C_{\infty_f}^* - C_{\infty_f}^* (1 - \exp(-K_L a_f \cdot t))) \qquad (6.36)$$

Hence,

$$\frac{dC}{dt} = K_L a_f \cdot C_{\infty_f}^* \cdot \exp(-K_L a_f \cdot t) \qquad (6.37)$$

Since Eq. (6.33) and Eq. (6.34) are the same, $(dC/dt = F)$, therefore,

$$F = K_L a_f \cdot C_{\infty_f}^* \cdot \exp(-K_L a_f \cdot t) \qquad (6.38)$$

Therefore,

$$F(t = 0) = K_L a_f \cdot C_{\infty_f}^* \qquad (6.39)$$

and $F_{(t=\infty)} = 0$.

Plotting this function F for gas depletion would give an exponential curve as shown in Figure 6.14. This is the baseline case plot. Without the action of microbial respiration, the oxygenation capacity of the aeration system is fully utilized. Eventually, the system will balance itself so that the tank becomes saturated, and the gas transfer is complete. Further continual supply of gas would not increase the oxygen content in the tank, and the system is said to be in a steady state, as the feed gas is balanced by the exit gas, and there is no gas depletion at steady state.

(ii) ASCE Model for R > 0

In the presence of cell respiration, according to current ASCE 18-96, Eq. (3.1), the gas depletion rate remains the same under the influence of R, but ASCE Eq. (2.2) now becomes:

$$\frac{dC}{dt} = K_L a_f (C_{\infty_f}^* - C) - R \qquad (6.40)$$

Integrating Eq. (6.40) with respect to time, gives

$$\frac{K_L a_f \cdot (C_{\infty_f}^* - C) - R}{K_L a_f (C_{\infty_f}^* - C_0) \quad R} = \exp(-K_L a_f \cdot t) \qquad (6.41)$$

Again, assuming $C_0 = 0$, at time $t = 0$, and re-arranging terms,

where
$$C = C_R (1 - \exp(-K_L a_f \cdot t)) \qquad (6.42)$$

From Eq. (6.40), $dC/dt = 0$ at steady-state, and $C = C_R$, then

$$C_R = C_{\infty_f}^* - \frac{R}{K_L a_f} \qquad (6.43)$$

Substituting Eq. (6.42) into Eq. (6.40) for C,

$$\frac{dC}{dt} = K_L a_f (C^*_{\infty_f} - C_R(1 - \exp(-K_L a_f \cdot t))) - R \tag{6.44}$$

Simplifying,

$$\frac{dC}{dt} = K_L a_f \cdot C_R \cdot \exp(-K_L a_f \cdot t) \tag{6.45}$$

From ASCE Eq. (3.1), or Eq. 6.34, where ASCE has been assumed to be the same as the baseline case,

$$F = K_L a_f (C^*_{\infty_f} - C) \tag{6.46}$$

Differentiating w.r.t. t,

$$\frac{dF}{dt} = -K_L a_f \cdot \frac{dC}{dt} \tag{6.47}$$

Substituting (6.45) into (6.47),

$$\frac{dF}{dt} = -K_L a_f \cdot K_L a_f \cdot C_R \cdot \exp(-K_L a_f \cdot t) \tag{6.48}$$

integrating,

$$F = K_L a_f \cdot C_R \cdot \exp(-K_L a_f \cdot t) + K \tag{6.49}$$

where K is an integration constant.

The boundary condition is that when $t \to \infty$, $F \to R$, and therefore $K = R$ Hence,

$$F = R + K_L a_f \cdot C_R \exp(-K_L a_f \cdot t) \tag{6.50}$$

But since from Eq. 6.43,

$$C_R = C^*_{\infty_f} - \frac{R}{K_L a_f}$$

Therefore,

$$F = R + (K_L a_f \cdot C^*_{\infty_f} - R)\exp(-K_L a_f \cdot t) \tag{6.51}$$

at $t = 0$, therefore,

$$F = K_L a_f \cdot C^*_{\infty_f}$$

at $t = \infty$, $F = R$, the following plot is obtained as shown in Fig. 6.15. The plot as shown in Figure 6.15 is similar to the baseline plot, except that the final steady state gas depletion rate at infinite time is not zero, but is given by the fixed respiration rate R. At steady state, therefore, the respiration rate equals the gas depletion rate which is concurrent with the thesis of this paper.

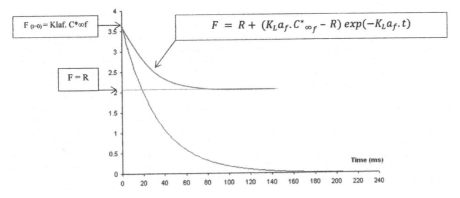

FIGURE 6.15 Gas depletion time variation $(R > 0)$ [ASCE model]

However, this plot as given by Figure 6.15 shows that the gas depletion is not impaired at the beginning in the presence of R. Like the previous plot for the case where cells are absent, the oxygenation capacity is fully utilized at time $t = 0$. Experiments have shown that this is not the case, and it is really not logical, since R must affect the gas depletion rate, no matter whether it is at the beginning, during, or at the end of the test. Mancy and Barlage (1968) described the phenomenon where long chain charged molecules attach to the gas bubble interfaces and impede the diffusion of oxygen to bulk solution. The longer the bubbles are in transit to the surface, the more these materials are attached to the bubbles resulting in a greater resistance to oxygen transfer and a reduction in alpha (α). Rosso and Stenstrom (2006b) have found that bubble surface contamination equilibrates even before detachment, so that after bubble detachment and during the transit of bubbles through the liquid, the liquid-side gas transfer coefficient K_L is reduced to a steady-state process value, *always lower than the gas transfer coefficient in pure water*. This means that the gas depletion must occur almost immediately upon detachment, and if the cells exert a transfer resistance, then the *reduction* of the gas-side depletion rate must start upon detachment at time $t = 0$, neglecting the bubble formation stage which is small compared to the time taken for the bubble transit to the surface. Therefore, the *gdp* at $t = 0$ must be smaller than the baseline case at $t = 0$. They should not be the same. This graph based on the ASCE model must therefore be incorrect.

(iii) Proposed Model for R > 0

Going through the same process, but with the ASCE Eq. (2.2) or Eq. 6.2 proposed to be changed to:

$$\frac{dC}{dt} = K_L a_f (C^*_{\infty_f} - C) - 2R \tag{6.52}$$

and the gas phase mass balanced is changed to:

$$F = K_L a_f (C^*_{\infty_f} - C) - R \tag{6.53}$$

the same expression for the gas depletion function is obtained, i.e.

$$F = R + K_L a_f \cdot C_R \exp(-K_L a_f \cdot t) \tag{6.54}$$

This is similar in expression as Eq. (6.43) above, but with C_R modified to:

$$C_R = C^*_{\infty f} - \frac{2R}{K_L a_f} \tag{6.55}$$

Therefore,

$$F = R + (K_L a_f \cdot C^*_{\infty f} - 2R) \exp(-K_L a_f \cdot t) \tag{6.56}$$

at $t = 0$, therefore,

$$F = K_L a_f \cdot C^*_{\infty f} - R \tag{6.57}$$

at $t = \infty$, $F = R$

The plot then becomes as shown in Figure 6.16.

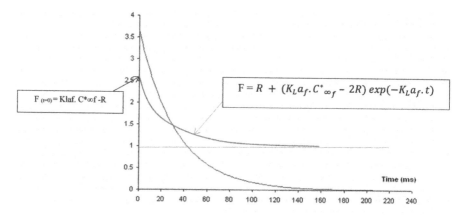

FIGURE 6.16 Gas depletion time variation ($R > 0$) modified model

This plot shows that the initial depletion rate is reduced by an amount equal to the respiration rate R. Experiments have borne out the fact that when respiring cells are present, the initial gas depletion must be smaller than when cells are absent, as evidenced by the higher off-gas content compared with the non-cell test. Furthermore, the non-cell condition would give a zero depletion rate at the end when the off-gas is equal to the feed gas content, whereas in the case of the oxygen uptake rate (OUR$_f$) R achieving a steady state, the off-gas mole fraction becomes constant at a lower value than 0.2095, and F at steady state equates to the respiration rate R.

Since the gas depletion represents the net oxygen transfer, the OTR$_f$ therefore equates to the consumption by the microbes, as is expected if a steady state is reached under the influence of the respiring cells. Using the principle of superposition, the *total* oxygen transfer rate remains given by $K_L a_f$ $(C^*_{\infty f} - C)$ as if the cells are not present (the baseline case), and $K_L a_f$ is then a fixed constant independent of R and gas depletion. This plot is therefore more correct for a consistent interpretation of the mass transfer coefficient $K_L a_f$.

The conclusion of this exercise is that, for submerged aeration where gas loss rate from the system is significant, the rate of transfer under the action of microbial respiration should be given by Eq. (6.52) reproduced below:

$$\frac{dC}{dt} = K_L a_f (C^*_{\infty f} - C) - 2R$$

This equation should then replace Eq. (2.2) in the ASCE 18-96 Guidelines. Experimental data does not exist to verify the gas depletion model as shown in Figure 6.16, since no data on direct comparison of a baseline case and a real case ($R > 0$) is available. However, since it is illogical to assume that R does not affect the gas depletion in the beginning of the test but does affect it at the end, it is likely the current ASCE model is incorrect.

Furthermore, if the same gas flow rate is applied, successive tests for estimating alpha using increasing MLSS (hence increasing steady state uptake rate) will indicate whether the initial gas depletion rate should be diminished by the uptake rate R.

This is indeed the case by examining Jing Hu's data [Hu 2006]. His data on the measurements of alpha and offgas values are plotted as shown in Figure 6.17 and Figure 6.18 (which is the same as Fig. 6.3). Figure 6.17 shows that there is a general trend of decreasing alpha; hence, in Figure 6.18, a decreasing gas depletion rate or increasing off-gas emission rate is obtained, for increasing MLSS or increasing R. In other words, the effect of R is a suppression of the gas depletion or a suppression of the net oxygen transfer rate in the system.

This phenomenon then agrees with the model that the gas depletion is given by Eq. 6.53 above:

$$F = K_L a_f (C^*_{\infty f} - C) - R$$

instead of the current ASCE 18-96 model for gas depletion rate given by ASCE's Eq. (3.1) as $F = K_L a_f (C^*_{\infty f} - C)$. Similarly, the gas transfer rate on the liquid phase would then be given by Eq. (6.52) above. It should be noted in passing that the above plot as shown in Figure 6.18 is obtained when the offgas data for the same test is plotted against the MLSS [Hu 2006].

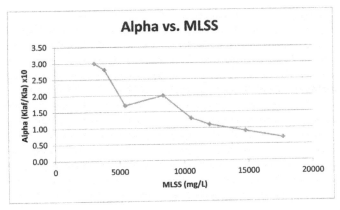

FIGURE 6.17 Relationship between alpha and MLSS for the membrane diffuser at 0.0283 m³/min (1 SCFM)

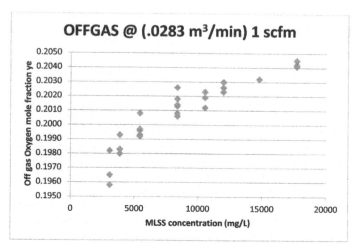

FIGURE **6.18** Relationship between offgas and MLSS for the membrane diffuser at 1 SCFM

6.6.4 A Simple Method to Eliminate the Impact of Free Surface Oxygen Transfer (Speculative)

McWhirter and Hutter (1989) points out that the mass transfer analysis of the oxygen transfer performance of diffused air or subsurface mechanical aeration systems has progressed very little over the past decades and is still true today. The recently-developed ASCE Standard [ASCE 2007] method for determination of the oxygen mass transfer performance of diffused or subsurface aeration systems is based on a greatly over-simplified mass transfer model. Although the *ASCE Standard* can be used to empirically evaluate point performance conditions, it does not provide a meaningful representation of the actual mass transfer process and is not capable of accurately assessing or predicting performance under changing operating or environmental conditions. According to DeMoyer et al. (2002), the standard testing methodology for oxygen transfer makes adjustment of measured values to other depths intangible.

Although they now have a model that separates the bubble mass transfer coefficient (K_La_b) and the surface transfer coefficient (K_La_s), their method only gives insight into the relative importance of transfer across the free water surface versus bubble surface. They rightly point out that bubble gas-water transfer is the dominant means of oxygen transfer. Their model results indicate that the surface transfer coefficient in a 9.25 m water depth circular tank of a diameter 7.6 m with an air flow rate of 51 to 76 scmh (U_g = .019 m/min to .028 m/min) is 59-85% of the bubble transfer coefficient. However, because of the hydrostatic pressure, the driving force inside a tank is higher than that on the surface, and so, the deeper the tank, the less is the relative importance of surface transfer. Also, the coefficients would depend on the gas flow rates.

The approach taken by this book is different from DeMoyer's approach in that the baseline mass transfer coefficient (K_La_0) is a lumped parameter that includes both effects. The model was developed based on first assuming that the surface transfer has no effect along with other assumptions such as constant bubble volume, and then later on adjusting the model by introducing calibration factors-n, m for the Lee-Baillod model, and e and $(1 - e)$ for the depth correction model. This approach has proven to be successful for translating the mass transfer coefficient K_La from one depth to another via the baseline K_La_0, within the cited range of gas flow rates tested, but is not expected to simulate well for high gas flow rate discharge and/or shallow tanks where gas transfer over the surface is expected to adopt a more prominent influence than the bubble transfer on which the theoretical development was based. This can be illustrated by citing some examples using the data obtained by GSEE, Inc. (available online at www.canadianpond.ca) who performed clean water tests at various air flow rates on a 6.5 m dia. circular tank and 9.45 m deep, using bubble tubings and OctoAir-10 aeration systems. Tests were carried out at 1.52 m (5 ft), 3.05 m (10 ft), 4.57 m (15 ft), and 6.10 m (20 ft) for the tubings. For the OctoAir, the depths ranged from 1.32 m, 2.84 m, 4.37 m to 5.89 m inclusive.

Using the method proposed in this book, the baseline coefficients were calculated and plotted against the average gas flowrates, as shown in Fig. 6.19 below. As can be seen, the curves are not equal, indicating that other factors are affecting the baseline curve. However, by assuming that the deepest tank has a baseline that is almost free from interference or suffering the least from such interference due to surface effect, the other tanks are then normalized to this tank with $Z_d = 4.57$ m. On examining the data, it can be shown that there is a definite correlation between the depth ratio of various tanks and the baseline coefficient and the correlation is a power function. The exponent for the depth ratio for the adjustment factor is found by minimizing the standard error between the predicted and measured baseline values, which will then give the best-fit values of the normalized baseline coefficients $K_La_0(N)$ as shown in Fig. 6.20.

The graphs (Fig. 6.20) show that the adjustment factor $(z/z)^{.5}$ appears to give a better correlation than $(z/z)^{.3}$, where $Z_d = 4.57$ m. The normalized baseline $K_La_0(N)$ for any average gas flow rate and tank height z_d would be given by Eq. 6.58 as stated below.

Similarly, for the OctoAir aeration device, it would appear that the baseline should be normalized to 5.89 m at a normalization factor of $(z/z)^{0.33}$, resulting in the graph shown in Fig. 6.21. (The last three points are for 4.37 m, 2.84 m, and 1.32 m, respectively).

$$K_La_0(N) = K_La_0 \times \text{sqrt}(z_d/Z_d) \qquad (6.58)$$

The contribution of the water surface transfer amounts to 9%, 21% and 39%, respectively. Once the normalization factor is determined, the baseline at any gas flow rate can be determined by Eq. 6.58, but with $(z/Z_d)^0.33$ as the correction factor instead of $(z/Z_d)^0.5$, the *normalized* baseline can then be used to simulate the mass transfer coefficient (K_La) for any other conditions just like the methodology used before for the other tests. It should be understood that the *normalized baseline* represents that which is free from the surface effect, so that where the gas discharge is large, or the tank is shallow, any significant surface transfer must be added to such simulation either afterwards, using an additional model such as the one postulated by DeMoyer et al. (2002), as a percentage of the estimated K_La, or by pro rata using Eq. 6.58 to calculate the true (surface + bubble) K_La_0 prior to the simulation. As an example, to calculate the standard baseline for the 1.32 m tank at a gas flow rate of 0.23 m^3/min, the reading from the graph gives a value of 0.01 min^{-1}, and so the true baseline mass transfer coefficient would be given by $(0.01) \times (5.89/1.32)^{0.33}$ which equals to 0.0164 min^{-1} for the baseline value. This value should then be used for any simulation on this tank for any gas flowrate and environmental conditions in order to calculate the standard mass transfer coefficient $(K_L a)_{20}$.

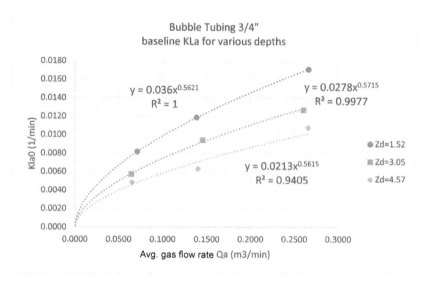

FIGURE 6.19 Relationship of baseline and gas flow rate before normalization

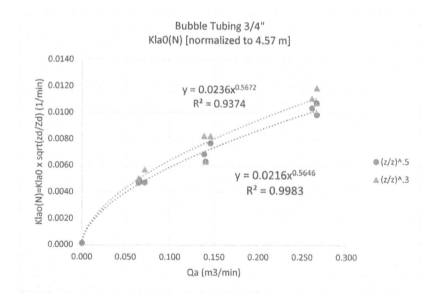

FIGURE 6.20 Comparison of two different normalization factors

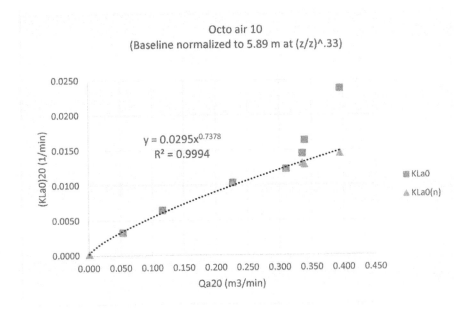

FIGURE 6.21 Relationship between the baseline and gas flow rate after normalization

6.7 NOTATION

α	wastewater correction factor, ratio of process water $\alpha \cdot K_L a$ to clean water $K_L a$
$\alpha \cdot F$	ratio of the process water $(\alpha \cdot K_L a)_{20}$ of fouled diffusers to the clean water $(K_L a)_{20}$ of clean diffusers at equivalent conditions (i.e. diffuser airflow, temperature, diffuser density, geometry, mixing, etc.), and assuming $(\alpha \cdot K_L a)_{20}/(K_L a)_{20} \approx (\alpha \cdot K_L a)/(K_L a)$
α'	wastewater correction factor, ratio of process water (with no microbes) $K_L a_f$ to clean water $K_L a$
β	correction factor for salinity and dissolved solids, ratio of $C^*_{\infty f}$ in wastewater to tap water
C^*_{∞}	oxygen saturation concentration in an aeration tank (mg/L)
$C^*_{\infty f}$	oxygen saturation concentration in an aeration tank under field conditions (mg/L)
$(C^*_{\infty})_{20}$	oxygen saturation concentration in an aeration tank at 20°C (mg/L)
$C^*_{\infty 0}$	hypothetical oxygen saturation concentration in an aeration tank at zero gas depletion rate (mg/L)
C_S	oxygen saturation concentration in an infinitesimally shallow tank, also known as oxygen solubility in water at test temperature and barometric pressure (mg/L)
C	dissolved oxygen concentration in a fully mixed aeration tank (mg/L)

C^*	a broad term for any saturation concentration in equilibrium with the partial pressure of oxygen in the gas phase, microscopically or macroscopically
C_R	the "apparent" or steady-state saturation concentration in the field (mg/L)
DO	Dissolved oxygen or dissolved oxygen concentration (mg/L)
$K_L a$	volumetric mass transfer coefficient, also known as the apparent volumetric mass transfer coefficient (min^{-1} or hr^{-1})
$(K_L a)_{20}$	volumetric mass transfer coefficient, also known as the apparent volumetric mass transfer coefficient at standard conditions (min^{-1} or hr^{-1})
$K_L a_f$	mass transfer coefficient as measured in the field, equals $\alpha' \cdot K_L a$ (min^{-1} or hr^{-1})
$K_L a_0$	baseline mass transfer coefficient, equivalent to that in an infinitesimally shallow tank with no gas side oxygen depletion (min^{-1} or hr^{-1})
$(K_L a_0)_{20}$	clean water baseline mass transfer coefficient at standard conditions, equivalent to that in an infinitesimally shallow tank with no gas side oxygen depletion at standard conditions (min^{-1} or hr^{-1})
$K_L a_{0f}$	wastewater baseline mass transfer coefficient, equivalent to that in an infinitesimally shallow tank with no gas side oxygen depletion (min^{-1} or hr^{-1})
$(K_L a_{0f})_{20}$	wastewater baseline mass transfer coefficient at standard conditions, equivalent to that in an infinitesimally shallow tank with no gas side oxygen depletion (min^{-1} or hr^{-1})
$\theta \cdot \tau \cdot \beta \cdot \Omega$	temperature, solubility, pressure correction factors as defined in ASCE 2007
MLSS	mixed liquor suspended solids (mg/L or g/L)
OTE	oxygen transfer efficiency (%)
p. SOTE	predicted standard oxygen transfer efficiency (%)
rpt. SOTE	reported standard oxygen transfer efficiency (%)
OTR	oxygen transfer rate (kg O_2/hr)
OTR_f	oxygen transfer rate in the field under process conditions, equals: $K_L a_f (C^*_{\infty f} - C)$ V-RV (for batch process) (kg O_2/hr)
OTR_{cw}	oxygen transfer rate under non-steady state oxygenationin clean water (kg O_2/hr)
OTR_{ww}	oxygen transfer rate under non-steady state oxygenationin wastewater without any microbial cell respiration (kg O_2/hr)
SOTR	standard oxygen transfer rate in clean water as defined in ASCE 2007 (kg O_2/hr)
OUR	oxygen uptake rate (kg O_2/hr)
OUR_f	oxygen uptake rate in the field, measurable by the off-gas method (kg O_2/hr)
R	microbial respiration rate, also known as the oxygen uptake rate (OUR) (kg O_2/m^3/hr usually expressed as mg/L/hr)

R_0	specific gas constant for oxygen (kJ/kg-K)
V	volume of aeration tank (m^3)
z	depth of water at any point in the tank measured from bottom, see Fig. 1 (m)
Z_d	submergence depth of the diffuser plant in an aeration tank (m)
Z_e	equilibrium depth at saturation measured from bottom, see Fig. 1 (m)
P_a, P_b	atmospheric pressure or barometric pressure at time of testing (kPa)
P_e	equilibrium pressure of the bulk liquid of an aeration tank (kPa) defined such that: $P_e = P_a + r_w d_e - P_{vt}$ where P_{vt} is the vapor pressure and r_w is the specific weight of water in N/m^3 (kPa)
d_e	effective saturation depth at infinite time (m)
Y_e	oxygen mole fraction at the effective saturation depth at infinite time
Y_0	initial oxygen mole fraction at diffuser depth, Y_d, also equal to exit gas mole fraction at saturation of the bulk liquid in the aeration tank, $Y_0 = 0.2095$ for air aeration
H	Henry's Law constant (mg/L/kPa) defined such that:

$$C_\infty^* = HY_e P_e \quad \text{or} \quad C_S = HY_0 P_a$$

Y_{ex}	exit gas or the off-gas oxygen mole fraction at any time
y	oxygen mole fraction at any time and space in an aeration tank defined by an oxygen mole fraction variation curve
gdp	gas-side oxygen depletion rate (equals OTR; equals zero at steady state for clean water; equals OTR$_f$ in process water; equals gdp_f at steady state) (kg O$_2$/hr)
gdp_{cw}	gas-side oxygen depletion rate in clean water (equals OTR$_{cw}$) (kg O$_2$/hr)
gdp_{ww}	gas-side oxygen depletion rate in wastewater (equals OTR$_{ww}$) (kg O$_2$/hr)
gdp_f	specific gas-side oxygen depletion rate due to microbes-induced resistance (equals OTR$_f$ at steady state); equals the microbial respiration rate R in process water (kg O$_2$/hr)
Q_a	height-averaged volumetric air flow rate (m^3/min or m^3/hr)
Q_S, AFR	gas (air) flow rate at standard conditions (20°C for US practice and 0°C for European practice), in (std ft^3/min or Nm^3/h)
n, m	calibration factors for the Lee-Baillod model equation for the oxygen mole fraction variation curve
W_{O2}	mass flow of oxygen in air stream (kg/h)

REFERENCES

American Society of Civil Engineers. (1997). Standard Guidelines for In-Process Oxygen Transfer Testing, 18-96. Reston, VA.

American Society of Civil Engineers. (2018). Standard Guidelines for In-Process Oxygen Transfer Testing. ASCE/EWRI 18-18. Reston, VA.

American Society of Civil Engineers. (2007). Standard 2-06: Measurement of oxygen transfer in clean water. Reston, VA.

Baillod, Robert C. (1979). Review of oxygen transfer model refinements and data interpretation. Proc., Workshop toward an Oxygen Transfer Standard, U.S. EPA/600-9-78-021, W.C. Boyle, ed., U.S. EPA, Cincinnati, 17-26.

Baquero-Rodríguez, G.A., Lara-Borrero, J. (2016). The influence of optic and polarographic dissolved oxygen sensors estimating the volumetric oxygen mass transfer coefficient (K_La). Mathematical Models and Methods in Applied Sciences 10(8): 142-151.

Baquero-Rodriguez, Gustavo A., Lara-Berrero, Jaime A., Nolasco, Daniel, Rosso, Diego. (2018). A critical review of the factors affecting modeling oxygen transfer by fine-pore diffusers in activated sludge. Water Environment Research 90(5): 431-441.

Bewtra, J., Tewari, P. (1982). Alpha and beta factors for domestic wastewater. Journal (Water Pollution Control Federation) 54(9): 1281-1287. Retrieved from http://www.jstor.org/stable/25041683.

Birima, A.H., Mohammed, T.A., Noor, M.J.M.M., Muyibi, S.A., Idris, A., Nagaoka, H., Ahmed, J., Ghani, L.A.A. (2009). Membrane fouling in a submerged membrane bioreactor treating high strength municipal wastewater. Desalination and Water Treatment 7(2009): 267-274.

Boon, Arthur G. (1979). Oxygen transfer in the activated-sludge process. Water Research Centre, Stevenage Laboratory, U.K. Proc., Workshop toward an Oxygen Transfer Standard, U.S. EPA/600-9-78-021, W.C. Boyle, ed., U.S. EPA, Cincinnati, 17-26.

Boyle, W.C., Campbell, H.J. Jr. (1984). Experiences with oxygen transfer testing of diffused air systems under process conditions. Water Science and Technology 16: 91-106.

Boyle, W.C., Stenstrom, M.K., Campbell, H.J. Jr., Brenner, R.C. (1989). Oxygen transfer in clean and process water for draft tube turbine aerators in total barrier oxidation ditches. Journal of the Water Pollution Control Federation, Vol. 61, No. 8, (Aug. 1989).

CEE 453: (2003). Laboratory Research in Environmental Engineering. http://ceeserver.cee.cornell.edu/mw24/Archive/03/cee453/Lab_Manual/pdf/Gas%20transfer.pdf.

Chiesa, S.C., Rieth, M.G., Ching, T. (1990). Evaluation of activated sludge oxygen uptake rate test procedures. Journal of Environmental Engineering, Vol. 116, No. 3, May/June, 1990.

DeMoyer, C.D., Schierholz, E.L., Gulliver, J.S., Wilhelms, S.C. 2002. Impact of bubble and free surface oxygen transfer on diffused aeration systems. Water Resarch 37(8): 1890-1904. DOI: 10.1016/S0043-1354(02)00566-3. PMID: 12697232.

Doyle, M. 1981. Small scale determination of alpha in a fine bubble diffuser system. Madison, WI: Department of Civil and Environmental Engineering, University of Wisconsin.

Eckenfelder, W.W. (1952). Aeration efficiency and design: I. Measurement of oxygen transfer efficiency. Sewage and Industrial Wastes 24(10): 1221-1228.

Eckenfelder, W.W. (1970). Water Pollution Control: Experimental Procedures for Process Design. Pemberton Press, Austin.

Garcia-Ochoa, F., Gomez, E. (2009). Bioreactor scale-up and oxygen transfer rate in microbial processes: an overview. Biotechnology Advances 27(2): 153-176.

Garcia-Ochoa, F., Gomez, E., Santos, V.E., Merchuk, J.C. (2010). Oxygen uptake rate in microbial processes: an overview. Biochemical Engineering Journal 49(3): 289-307.

Hu, J. (2006). Evaluation of parameters influencing oxygen transfer efficiency in a membrane bioreactor (Doctoral dissertation).

Hwang, H.J., Stenstrom, M.K. (1985). Evaluation of fine-bubble alpha factors in near full-scale equipment. Journal (Water Pollution Control Federation), Volume 57, Number 12.

Jiang, L.M., Garrido-Baserba, M., Nolasco, D., Al-Omari, A., DeClippeleir, H., Murthy, S., Rosso, D. (2017). Modelling oxygen transfer using dynamic alpha factors. Water Research 124: 139-148. doi: 10.1016/j.watres.2017.07.032. PMID: 28753496.

Jiang, P., Stenstrom, M.K. (2012). Oxygen transfer parameter estimation: impact of methodology. Journal of Environmental Engineering 138(2): 137-142.

Lee, J. (2017). Development of a model to determine mass transfer coefficient and oxygen solubility in bioreactors. Heliyon 3(2): e00248.

Lee, J. (2018). Development of a model to determine the baseline mass transfer coefficients in aeration tanks. Water Environment Research 90(12): 2126.

Lewis, W.K., Whitman, W.G. (1924). Principles of gas absorption. Industrial and Engineering Chemistry 16(12): 1215-1220.

Mancy, K.H., Barlage, W.E., Jr. (1968). Mechanism of interference of surface active agents with gas transfer in aeration systems. pp. 262-286. *In*: Gloyna F., Eckenfelder, W.W. Jr. (eds.). Advances in Water Quality. University of Texas Press, Austin, TX.

Mahendraker, V. (2003). Development of a unified theory of oxygen transfer in activated sludge processes – the concept of net respiration rate flux. Department of Civil Engineering, University of British Columbia.

Mahendraker, V., Mavinic, D.S., Rabinowitz, B. (2005a). Comparison of oxygen transfer test parameters from four testing methods in three activated sludge processes. Water Qualilty Resource Journal Canada 40(2): 164-176.

Mahendraker, V., Mavinic, D.S., Rabinowitz, B. (2005b). A simple method to estimate the contribution of biological floc and reactor-solution to mass transfer of oxygen in activated sludge processes. Wiley Periodicals, Inc. DOI: 10.1002/bit.20515.

McKinney, Ross E., Stukenberg, John R. (1979). On-site Evaluation: Steady State vs. Non-Steady State Testing. Proc., Workshop toward an Oxygen Transfer Standard, U.S. EPA/600-9-78-021, W.C. Boyle, ed., U.S. EPA, Cincinnati, 17-26.

McCarthy, J. (1982). Technology Assessment of Fine Bubble Aerators. U.S. Environmental Protection Agency, Washington, D.C., EPA/600/2-82/003 (NTIS PB82237751).

McWhirter, John R., Hutter, Joseph C. (1989). Improved oxygen mass transfer modeling for diffused/subsurface aeration systems. American Institute of Chemical Engineers Journal 35(9): 1527-1534. https://doi.org/10.1002/aic.690350913.

Metcalf & Eddy. (1985). Wastewater Engineering: Treatment, Disposal, Reuse, 2nd ed. McGraw-Hill, New York.

Mott, H.V. (2013). Environmental Process Analysis: Principles and Modelling. John Wiley & Sons, Canada.

Mueller, J.A., Boyle, W.C. 1988. Oxygen transfer under process conditions. Journal Water Pollution Control Federation 60: 332-341.

Rosso, D., Stenstrom, M. (2006a). Alpha Factors in Full-scale wastewater aeration systems. Water Environment Foundation. WEFTEC 06.

Rosso, D., Stenstrom, M. (2006b). Surfactant effects on α-factors in aeration systems. Water Research 40(7): 1397-1404, Elsevier Ltd.

Santos, Mazorra, V.E., Galdeano, C., Gomez, E., Alcon, A., Garcia-Ochoa, F. (2006). Oxygen uptake rate measurements both by the dynamic method and during the process growth of *Rhodococcus erythropolis* IGTS8: modelling and difference in results. Biochemical Engineering Journal 32(2006): 198-204.

Shraddha, M., Narang, Atul. (2018). Quantifying the parametric sensitivity of ethanol production by Scheffersomyces (Pichia) stipitus: development and verification of a method based on the principles of growth on mixtures of complementary substrates. Microbiology 164(11): 1348-1360. 10.1099/mic. 0.000719.

Stenstrom, Michael K., Gilbert, R. Gary. (1981). Effects of alpha, beta and theta factor upon the design, specification and operation of aeration systems. Water Research 15(6): 643-654.

Tchobanoglous, G., Burton, F.L., Stensel, H.D., Eddy, M. (2003). Wastewater Engineering: Treatment and Reuse, McGraw-Hill Series in Civil and Environmental Engineering, Fourth ed. McGraw-Hill New York, NY.

Yunt, Fred W., Hancuff, Tim O., Brenner, Richard C. (1988a). EPA/600/2-88/022. Aeration equipment evaluation. Phase 1: Clean water test results. Contract No. 14-12-150. Los Angeles County Sanitation District, Los Angeles, CA. Municipal Environmental Research Laboratory Office of Research and Development, U.S. EPA, Cincinnati, OH.

Yunt, Fred W., Hancuff, Tim O. (1988b). EPA/600/S2-88/022. Project summary: aeration equipment evaluation. Phase I: Clean water test results. Water Engineering Research Laboratory, Cincinnati, OH.

Yunt, Fred W., Stenstrom, Michael K. (1988c). Aeration equipment evaluation – Phase II Process Water Test Results. Contract No. 68-03-2906. (undated report). Los Angeles County Sanitation Districts, Risk Reduction Engineering Laboratory Office of Research and Development, U.S. EPA, Cincinnati, OH.

Zhou, Xiaohong, Wu, Yuanyuan, Shi, Hanchang, Song, Yanqing. (2012). Evaluation of oxygen transfer parameters of fine-bubble aeration system in plug flow aeration tank of wastewater treatment plant. Journal of Environmental Sciences 25(2): 295-301.

7

Recommendation for Further Testing and Research

7.0 INTRODUCTION

The alpha factor (α) represents the ratio of the mass transfer coefficient in process water $K_L a_f$ to $K_L a$ in clean water at equivalent test conditions and this ratio can range from approximately 0.1 to greater than 1.0 (ASCE 2007). The subscript f signifies field conditions. This wide range makes it very difficult to design an aeration tank for a wastewater treatment plant. An opportunity exists for testing for relationships between alpha (α) and tank volume. The Water Research Foundation (WRF) has accepted a pre-proposal for initiating such a project, based on new model discoveries in recently published papers [Lee 2017, 2018]. If accepted, the foundation would sponsor \$75,000. The budgeted project would require the erection of a tank, say 4.6 m (15 ft) to 6.1 m (20 ft) tall, and with a tentative plan area of 2 m by 3 m. The proposal applies to diffused bubble aeration only.

Other investigators have concluded that it is not alpha versus tank volume that matters, but alpha versus diffuser submergence [Boon and Lister 1973, 1975][Doyle et al. 1983] [Groves et al. 1992] [Rosso and Stenstrom 2006a, b]. In general, these researchers found that as diffuser submergence is increased, alpha is reduced. Mancy and Barladge (1968) described the phenomenon where long chain charged molecules attach to the gas bubble interfaces and impede the diffusion of oxygen to the bulk aqueous solution as dissolved oxygen (DO). The longer the bubbles are in transit to the free water surface, the more these materials are attached to the bubbles, resulting in a greater resistance to oxygen transfer and a reduction in alpha. However, Keil and Russell (1987) developed the bubble recirculation cell to model sparged aeration tank. The method regards each sparger (or diffuser) to be substantially independent of neighboring spargers, and hence the aerated liquid can be divided into cells, each cell corresponding to the space around one sparger. This approach means that the behavior of a single sparger can be studied and the results then can be applied to the whole aerated volume if the bulk liquid is completely mixed. This is especially true if the diffuser plants are arranged as full-floor coverage.

In clean water aeration, the proposed new model called the Lee-Baillod model allows plotting the mole fraction curve along the tank height starting from the diffuser submergence depth Z_d. At equilibrium, the mole fraction is at a minimum y_e which occurs at an effective depth d_e [ASCE 2007], as shown in Fig. 7.1 below.

Furthermore, another proposed model, called the depth correction model [Lee 2018], then allows plotting $K_L a$ with depths based on a baseline $K_L a$ that is defined as the mass transfer coefficient at zero depth, essentially meaning zero gas depletion.

The author postulates that only then can one compare clean water K_La with in-process water K_La_f that would make alpha invariant. Since the model depends on the superficial gas flow rate (U_g) that depends on the horizontal cross-sectional area of the tank for a fixed gas supply, the volume does come into play.

7.1   TESTING FOR ALPHA IN AERATION TANK

After building a tank of an appropriate size, it may then be possible to see if a bench scale experiment can actually predict the K_La in a higher scale. The work so far has agreed that K_La would decrease with depth which is why the baseline K_La is always the maximum value. By testing at different depths and different flow rates, it may, according to the proposed thesis, yield a *constant* baseline value. If this is true, then rating curves may be readily produced for aeration systems for various gas flow rates, depths, and other environmental conditions. Some researchers believe that the baseline K_La at zero depth is a hypothetical concept that has no practical value, and have suggested that study must be carried out on full-scale systems. Then there is no need to worry about scale up. According to them, in real systems there are so many things that impact alpha, including sludge retention time (SRT) and organic loading among others, which are so highly variable that generating meaningful rating curves is not deemed possible. The proposer disagrees with the above argument that the baseline has no practical value. This proposal has included a worked example of how to use the baseline to predict or simulate mass transfer coefficients for full scale plants. The proposal is that these other 'things' can be separately modelled, so that alpha pertains only to the waste characteristics as an intensive property, while the mass transfer equations are modified to include the other things as extensive properties. The baseline K_La is a hypothetical parameter defined mathematically as

$$\lim_{Zd \to 0} K_La = K_La_0, \tag{7.0}$$

where K_La_0 represents the baseline. Since every tank under aeration has gas depletion in submerged aeration, it is not possible to relate tanks of different height or diffuser submergence (Z_d), unlessthey are all reduced to zero depth or to a very, very shallow depth as to be infinitesimal. The proposed Lee-Baillod model relates K_La to depths based on this hypothetical parameter. For every tank, it is possible to back calculate from the measured K_La to this baseline where the gas depletion becomes zero or approaching zero. Baillod [1979] called this the "true" K_La [see Appendix].

Since every tank no matter how shallow has some physical height, there is no such thing as "true" K_La in any clean water test. However, when extrapolating to zero depth, it becomes a baseline, from which other depths can be contemplated. Based on Yunt et al.'s experiments [Yunt et al. 1980] [Yunt 1988], tanks of 3.05 m (10 ft), 4.57 m (15 ft), 6.09 m (20 ft) and 7.62 m (25 ft) were measured. The proposer found that each of these tanks can be back calculated to find the baseline and the baseline is a constant no matter what the tank depth is, when the gas flow rate is normalized to the same "height-averaged" volumetric flow rate. The error of estimation is around 1 ~ 3%.

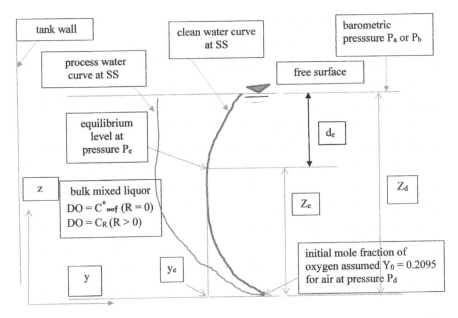

FIGURE 7.1 Oxygen mole fraction curves at saturation for in-process water compared to clean water based on the Lee-Baillod model (e = equilibrium; SS = steady state) [Lee 2018]

Therefore, it is hypothesized that the same model would apply to wastewater under aeration (as shown by the second curve, for process water, in Fig. 7.1). At steady state, the initial mole fraction curve is shifted to the left due to the effect of microbial activities. How much it is shifted would depend on the respiration rate of the microbes. The proposer postulates that this amount of shifting of the oxygen mole fraction curve can be estimated by the *principle of superposition* [Lee 2018, 2019b] by first assuming that the microbiological effect is negligible, and then vectorially adding the effect to the mass balance equation. If one can determine $K_L a_f$ in bench scale or pilot scale, it should be possible to calculate the corresponding $K_L a_f$ in full scale, based on the depth correction model verified by clean water tests. This is the hypothesis underlying this proposal. The parameter alpha (α) is normally measured in the field as a common practice. Such measurement necessarily includes all the effects of field variables on the mass transfer coefficient. Such an approach confounds the meaning of alpha with too many variables. When Eckenfelder (1952) first designed his test, the biological solids were filtered out to exclude the biological interference, so that his alpha value is dependent only on the wastewater characteristics which is, for all intents and purposes, an intensive property (i.e. independent of scale) of the mixed liquor. When the wastewater characteristics are constant, alpha is constant.

It is possible that the magnitude of $K_L a$ decreases in systems where there is biological growth. According to some researchers, that could be because, for instance, a high rate of O_2 utilization means a lot of suspended bacteria in the water, and that affects the aqueous diffusivity of O_2 in the water (exactly how is not known, since suspended bacteria only constitute a tiny fraction of the bulk mass of the liquid). Hence, the "individual" mass-transfer coefficients k_L and k_G could be affected. Since $(1/K_L) = (1/k_L) + (1/H * k_G)$, so if the chemistry or physics of the water decreases k_L, then K_L will also be decreased. This could explain why mass transfer is lower than expected in systems with biological activity. But this is still because of the changes in wastewater characteristics (by whatever means) that changes the diffusivity, resulting in a change of resistance to gas transfer. Therefore, by relating alpha only to the wastewater characteristics, it could at least eliminate one variable, the biological activities affecting alpha that leave the other variables to be elucidated one by one by other means. This alpha should be distinguished by a different symbol such as α' to avoid confusion with the current understanding of alpha.

Mahendraker et al. (2005) postulated that the overall resistance in the mixed liquor is the sum of the liquid's resistance (i.e. due to the water characteristics alone) and an additional resistance due to the biological floc. This concept leads to the equation that $1/\alpha K_L a = 1/\alpha' K_L a + 1/K_L a_{bf}$ where the subscript bf denotes biological floc which is a function of the respiration rate R. In their paper, the authors did not correlate the biological floc with the respiring rate R, but it can be seen that the parameter $K_L a_{bf}$ has the same form as in the basic transfer equation. From this, it can be shown that $\alpha' = 2\alpha$ which is similar to the hypothesis proposed by the author whose proposed understanding of alpha is α'. This has been the subject of another paper [Lee 2019b]. If this concept is true, Eq. CG-1 in ASCE 2-06 [ASCE 2007] would become modified to the following:

$$OTR_f = \left(\frac{1}{(C_\infty^*)_{20}} \right) [\alpha (SOTR)\theta^{T-20}](\tau \cdot \beta \cdot \Omega \cdot (C_\infty^*)_{20} - C) - RV \qquad (7.1)$$

where $\alpha = \alpha'$ and other symbols are as defined in the Standard [ASCE 2007]. This equation would require determination of the respiration rate R as opposed to the original equation where the effect of biological activities is included in alpha [Lee 2019b].

7.2   PROPOSED TEST FACILITY

Most clean water testing is performed using the clean water standard developed by the American Society of Civil Engineers (ASCE 2007). It is proposed that an outdoor, all-steel, rectangular aeration tank with dimensions of 6.1 m × 6.1 m × 7.6 m sidewater depth (SWD) should be used for all tests. Depending on budget, the plan size of the tank may be reduced, perhaps to half-scale. The test tank for use in this project is proposed to be similar to that shown in Fig. 7.4 and Fig. 7.5 below extracted from the literature [Wagner et al. 2008]. To study the wall effect, a small scale test column to the same maximum height can be separately erected to serve as a control. Potable water is to be used in all clean water tests for this study. The air delivery system and other equipment to be used will be similar to those used in Yunt's test program [Yunt et al. 1980], and tests are to be performed in accordance with the ASCE standard as far as possible. The following flowchart procedure lays out the steps:

- First, clean water tests (CWT) are to be done for 2 ~ 3 temperatures, preferably one below 20°C, one at 20°C and one above. CWT is also required for 2 different gas flow rates, so that altogether a minimum of 4 tests are recommended for a tank of adequate size, and the tank water depth is suggested to be fixed at 3 m (10 feet) or 5 m (15 feet) or any other depth of choice. Tests will be repeated several times to have a constant $K_L a$ for each temperature, and the test is to be repeated for different applied gas flow rates so that the $K_L a$ vs. Q_a relationship can be estimated.
- All diffused aeration systems will experience gas-side depletion as the water depth increases. This change in gas-side depletion is dealt with by the Lee-Baillod model, allowing calculation of the baseline $(K_L a_0)$ using the Microsoft Excel Solver or similar where $K_L a_0$ is a variable to be determined, with the measured $K_L a$ and C_∞^* as the independent variables.
- Once the baseline $K_L a_0$ is established, a specific standard baseline can be determined using the temperature correction model and the established $K_L a_0$ vs. Q_a relationship, and this value can be used to find the transfer coefficient at another tank depth. The Excel Solver or similar is used to solve the simultaneous equations, using the established baseline parameter $K_L a_0$, as well as the actual environmental conditions surrounding the scaled-up tank. Hence, the same Solver method is used twice, to calculate both the $K_L a_0$ and the $K_L a$. The temperature correction model of choice is the 5th power model [Lee 2017] as advocated by the author.
- All the measured apparent $K_L a$ values can be used to formulate the relationship between $K_L a$ and Q_a, but the resultant slopes may have some differences. These should be compared to the plot of $(K_L a_0)_T$ vs. Q_{aT} and also with $(K_L a_0)_{20}$ vs. Q_{a20}. The latter curve should give the best correlation. Likewise, all $(K_L a_0)_T$ values are to be plotted against their respective handbook solubilities $(C_S)_T$.
- The specific baseline $(K_L a_0)_{20}/Q_{a20}{}^q$ [Zhou et al. 2012] [Lee 2018] is expected to be constant for all the tanks tested. From the standardized baseline $(K_L a_0)_{20}$ at 20°C, a family of rating curves for the standard mass transfer coefficient $(K_L a)_{20}$ can thus be constructed for various gas flow rates applied to various tank depths using Eqs. (3.6 to 3.10) as stated in Chapter 3.

7.3 TECHNICAL CHALLENGES

Presumably, changes in tank shapes and sizes, diffuser layouts and tank depths will affect the oxygen transfer film and the K_L value of the scaled-up tank. Since the model is a holistic approach on the overall change in both parameters K_L and 'a', one of the challenges is in determining the limitations and the validity boundaries of this model. A trial-and-error approach to establish the boundaries may be necessary. The proposed model appears correct and valid for the set of tanks cited in the paper [Lee 2018] [Yunt et al. 1980], but the goal of scaling up test data from a smaller tank size to a larger tank size may depend on external factors that may confound the new model. Transfer devices typically produce irregularly sized bubbles that often swarm in various hydrodynamic patterns, e.g. spiral roll devices vs. full-floor coverage. So, scale-up changes that effect on the size, shape, depth and roll patterns in a tank effect on oxygen transfer performance. Small tanks are notorious for wall effects. The second challenge would be in the selection of suitable size and shape of the test tank that would give the best geometric similitudes between the various tests. It would also be important to select the aeration system that would not produce excessive plume entrainment of air during the testing [DeMoyer et al. 2002].

It is difficult to describe a required geometry or placement for testing conducted in tanks other than the full-scale field facility. According to the ASCE Standard, appropriate configurations for shop tests should simulate the field conditions as closely as possible. For example, width-to-depth or length-to-width ratios should be similar. Potential interference resulting from wall effects and any extraneous piping or other materials in the tank should be minimized. The density of the aerator placement, air flow per unit volume, or area and power input per unit volume are examples of parameters that can be used to assist in making comparative evaluations. However, the work here is to prove that, for the same configurations of aerator placement and tank dimensions, the model is able to predict oxygen transfer efficiency for a range of tank water depths using a universal standard specific baseline mass transfer coefficient $(K_L a_0)_{20}$.

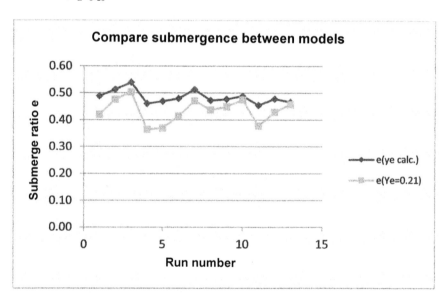

FIGURE 7.2 Comparison of effective depth ratio: rigorous analysis versus ASCE method

7.4   ESTIMATION OF THE EFFECTIVE DEPTH RATIO $(e = d_e/Z_d)$

Figure 7.2 (also Fig. 3.2) is a plot of the effective depth ratios calculated from the test runs for the FMC diffuser tests, the lower line showing the results based on a constant equilibrium mole fraction at 0.21 similar to the equation in ASCE 2-06 Annex F [ASCE 2007], while the top line was based on the developed model equations which describe the mole fraction variation curve andthe calibration parameters, n and m. It is important to note that the ASCE 2-06 Annex F Eq. (F-1) has treated Y_e to be the same as Y_0 which is not correct. As a result of this rigorous analysis, the top line in Fig. 7.2 gives a more consistently uniform depth ratio of $e = d_e/Z_d$.

It is to be noted the following for the development of the effective depth evaluation based on ye:

The English chemist <u>William Henry</u>, who studied the topic of gas solubility in the early 19th century, in his publication about the quantity of gases absorbed by water, described the results of his experiments:

"... water takes up, of gas condensed by one, two, or more additional atmospheres, a quantity which, ordinarily compressed, would be equal to twice, thrice, &c. the volume absorbed under the common pressure of the atmosphere."

Unfortunately, Henry killed himself in the end, perhaps because his brilliancy was not fully appreciated. If he hadn't died young, he might have discovered more, such as Dalton's Law, or the use of his law together with Dalton's Law.

Dalton's Law says that in a mixture of gases within a vessel, the total pressure is the sum of the individual partial pressures of the gases inside the vessel.

Combining the two physical laws, we have

$$C_{st} = H \cdot (Y \cdot P) \tag{7.2}$$

where H is Henry's law constant; C_{st} is the gas solubility. In the standard, $C_{st} =$ tabular value DO surface saturation concentration at test temperature, standard total pressure of 1.00 atm (101.3 kPa) and 100% relative humidity, in mg/L; $Y =$ oxygen mole fraction in the gas phase; the product Y.P is the partial pressure in equilibrium with the liquid phase, where P is the total pressure.

Henry's law can be applied to any point within a bulk liquid. When applied to the surficial situation under atmospheric pressure P_a, the partial pressure of oxygen is $0.21 \cdot P_a$; hence, the law constant can be determined by re-arranging the above equation to form:

$H = C_{st}/(0.21 \cdot P_s)$ where P_s is standard barometric pressure, 101.3 kPa, in kPa (atm).

Now, the vapor pressure has an effect on solubility so that a correction needs to be made to the surficial pressure so that

$$H = C_{st}/(0.21 \cdot (P_S - P_{vt})) \tag{7.3}$$

Now, Henry's law can also be applied to the equilibrium point for a bulk liquid under aeration, so that

$$C_\infty^* = H \cdot (Y_e \cdot P_e) \tag{7.4}$$

Substituting H by the previous equation, we have

$$C_\infty^* = [C_{st}/(0.21 \cdot (P_S - P_{vt}))] (Y_e \cdot P_e) \tag{7.5}$$

It is unfortunate that the ASCE standard has assumed $Y_e = 0.21$ as well, so that in the standard, the mole fraction is cancelled out, giving:

$$C_\infty^* = C_{st}/(P_S - P_{vt}) \cdot P_e \tag{7.6}$$

The equilibrium pressure, if we assume vapor pressure has similar effect, is given by: $P_e = P_a + rw \cdot de - P_{vt}$ where P_a is at test condition, also symbolized by P_b, the barometric pressure.

Therefore,

$$C_\infty^* = C_{st}/(P_S - P_{vt}) \cdot (P_a + rw \cdot de - P_{vt}) \tag{7.7}$$

Hence,

$$de = 1/rw \left[(C_\infty^*/C_{st}) \cdot (P_S - P_{vt}) - P_a + P_{vt} \right] \tag{7.8}$$

which is identical to the ASCE 2007 Annex F equation for the estimation of d_e.

Unfortunately, this equation is wrong or approximate only-the mole fraction of oxygen in the bubble at the equilibrium point cannot be ignored! Although oxygen is only slightly soluble in water, the mole fraction of oxygen in the bubble at the equilibrium point is calculated by the gas-side gas depletion following a mole fraction variation curve, and at the equilibrium point the mole fraction is different from 0.21. It could be more, it could be less, depending on the initial gas composition at the point of release from the diffuser and the other factors such as gas depletion. In ordinary circumstances, it is usually slightly less, so that $Y_e < 0.21$. Therefore,

$$C_\infty^* = [C_{st}/(0.21 \cdot (P_S - P_{vt}))] \cdot Y_e \cdot P_e$$

Rearranging gives

$$de = 1/rw \left[(C_\infty^*/C_{st}) \cdot (P_S - P_{vt}) \cdot 0.2095/Y_e - P_a + P_{vt} \right] \tag{7.9}$$

Equation 7.9 is an additional equation to the original five developed equations (Eq. 3.6–Eq. 3.10) relating the effective depth, de, to the oxygen mole fraction at equilibrium, Y_e, for simulation giving a total of six equations with six unknowns (Y_e, n, m, de, $K_L a$, C_∞^*) when the baseline $K_L a_0$ is an independent input variable pre-determined by clean water tests.

7.5 STANDARD OXYGEN TRANSFER RATE

In accordance with the ASCE 2-06 standard [ASCE 2007], the average value of SOTR shall be calculated by averaging the values at each of the n determination points by

$$\text{SOTR} = \frac{V}{n} \sum_{i=1}^{n} K_L B_{20i} C_{\infty 20i}^* \tag{7.10}$$

$$\text{SOTR} = \frac{1}{n} \sum_{i=1}^{n} \text{SOTR}_i \tag{7.11}$$

where

$$\text{SOTR}_i = K_L B_{20i} C_{\infty 20i}^* V \tag{7.12}$$

The following photographs (Fig. 7.3, Fig. 7.4) extracted from the literature are an example of a test tank suitable for the purpose of this proposal:

FIGURE **7.3** Schematic of proposed test tank [Wagner et al. 2008]

7.6   STANDARD OXYGEN TRANSFER EFFICIENCY (SOTE)

Oxygen transfer efficiency refers to the fraction of oxygen in an injected gas stream dissolved under given conditions. The standard oxygen transfer efficiency (SOTE), which refers to the OTE at a given gas rate (see ASCE 2007 Annex A), water temperature of 20°C, and barometric pressure of 1.00 atm (101.3 kPa), may be calculated for a given flow rate of air by:

$$\text{SOTE} = \text{SOTR}/W_{02} \qquad (7.13)$$

where

SOTE = standard transfer efficiency as a fraction; and

W_{02} = mass flow of oxygen in air stream, lb/min (kg/hr)

For subsurface gas injections systems, the value of SOTE should be reported as per ASCE 2007 Section 8.4. If possible, the standard deviations of the parameter estimates, $K_L a$, C_∞^*, and standard error of estimate should also be reported. The above applies to clean water. Under process conditions, the *principle of conservation of mass* must be applied, so that the OTE_f is given by OTR_f/W_{02} with the same units as for clean water test, where OTR_f is given by Eq. 7.1, and the Standard Model would be written as:

$$\text{OTR}_f = V\frac{dC}{dt} + RV \qquad (7.14)$$

where OTR_f is given by Eq. 7.1 which is based on the *principle of superposition* equivalent to:

$$OTR_f = K_L a_f (C_{\infty f}^* - C)V - RV \qquad (7.15)$$

Equating the two equations for the OTR_f gives:

$$K_L a_f = \frac{2R + \dfrac{dC}{dt}}{C_{\infty f}^* - C} \qquad (7.16)$$

Therefore, the above equations show that the mass transfer coefficient can be determined in the field provided that the respiration rate R is known.

FIGURE 7.4 Proposed diffused aeration configuration (diffusers can be soaker hose pipes or similar)

7.7 DETERMINATION OF OXYGEN TRANSFER IN WASTEWATER

7.7.1 Procedure for Determination of Dissolved Oxygen Uptake Rate and Oxygen Transfer Rate

In principle, the determination of the oxygen transfer rate (OTR) in an aerobic bioreactor is remarkably simple. At any point in time during fermentation or wastewater treatment, the oxygen uptake rate (OUR) must equal the transfer rate when the process is at a steady state. At steady state, there is no change of dissolved oxygen (DO) concentration with time so that all the oxygen transferred is uptaken by the microbes within the bioreactor, provided that the content is well-mixed and both the microbes and the DO are uniformly and spatially distributed within the reactor vessel. At steady state, there is no oxygen uptake by the aqueous solution or the reactor solution, provided no chemical oxidation processes are occurring. Therefore, the mass balance equation is simply given by:

$$OTR = OUR \qquad (7.17)$$

The difficulty is that there is no direct method to measure either the oxygen transfer rate or the oxygen uptake rate, resulting in only an approximate method of evaluation.

Traditionally, the OTR determination relies on the mass transfer coefficient ($K_L a$) which is an uncertain parameter. The determination is based on a non-steady state method as per ASCE 2-06 standard or similar, which is an indirect method. On the other hand, the OUR is also subject to many uncertainties, so that discrepancies can be as much as 50% even under the best circumstances when attempts are made to equate OUR with OTR. The various methods to be investigated for determining the OUR are described below:

7.7.2   BOD Bottle Method

The proposer reckons there are only three possible explanations for the increase of the OUR using this method, compared to the actual OUR *in-situ*. Firstly, when the sample is agitated by vigorous shaking, the activity level of the microbes might increase and so they respire more. Secondly, the oxygen level in the sample may be limiting (although at 2 p.p.m., it shouldn't be) and so the respiration rate is not as high as in a normal DO level. When the DO is elevated artificially to a higher level, the oxygen availability is increased giving a higher respiration rate in the sample; thirdly, Garcia-Ochoa et al. (2010) suggested a cell economy principle by which the microbes voluntarily reduce the respiration level at low DO or at declining DO from a high level by quorum sensing or otherwise, changing the respiration rate at elevated level due to increased oxygen availability and conversely, reducing the respiration rate at decreasing DO concentration. There have been many literatures on this but none seems to have given a definite answer. Chisea et al. [1990] conducted a series of bench- and pilot-scale experiments to evaluate the ability of biochemical oxygen demand (BOD) bottle- based oxygen uptake rate (OUR) analyses to represent accurately *in-situ* OUR in complete mix-activated sludge systems. Aeration basin off-gas analyses indicated that, depending on system operating conditions, BOD bottle-based analyses could either underestimate *in-situ* OUR rates by as much as 58% or overestimate *in-situ* rates by up to 285%. A continuous flow respirometer system was used to verify the off-gas analysis observations and assessed better the rate of change in OUR after mixed liquor samples were suddenly isolated from their normally continuous source of feed. OUR rates for sludge samples maintained in the completely mixed bench-scale respirometer decreased by as much as 42% in less than two minutes after feeding was stopped. However, this limitation can be somewhat addressed by maintaining feed during the sample analysis. *Based on these results, BOD bottle-based OUR results should not be used in any complete mix-activated sludge process operational control strategy, process mass balance, or system evaluation procedure requiring absolute accuracy of OUR values. It should also be noted that the intensive aeration in the respirometer, which was rather violent and may have broken up the floc, caused increased delivery of DO and substrate to the floc, so that the respirometer method is not considered reliable as well [Doyle et al. 1983].*

7.7.3   Synthetic Wastewater to Determine Actual Oxygen Uptake Rate (Mines' Method)

Mines et al. (2016) indicated that the proper design of aeration systems for bioreactors is critical since it can represent up to 50% of the operational and capital cost at water reclamation facilities. Transferring the actual amount of oxygen needed to meet the oxygen demand of the wastewater requires α- and β-factors, which are used for calculating the actual oxygen transfer rate (AOTR) under process conditions based on the standard oxygen transfer rate (SOTR). In their experiment, the SOTR is measured in tap water at 20°C, 1 atmospheric pressure, and 0 mg L^{-1} of dissolved oxygen (DO). In their investigation, two 11.4-L bench-scale completely mixed activated process (CMAS) reactors were operated at various solid retention times (SRTs) to ascertain the relationship between the α-factor and SRT, and between the β-factor and SRT. The second goal was to determine if actual oxygen uptake rates (AOURs) are equal to calculated oxygen uptake rates (COURs) based on mass balances. Each reactor was supplied with 0.84 L m^{-1} of air resulting in SOTRs of 14.3 and 11.5 g O_2 d^{-1} for Reactor 1 (R-1) and Reactor 2 (R-2), respectively. The estimated theoretical oxygen demands of the synthetic feed to R-1 and R-2 were 6.3 and 21.9 g O_2 d^{-1}, respectively. R-2 was primarily operated under a dissolved oxygen (DO) limitation and high nitrogen loading to determine if nitrification would be inhibited from a nitrite buildup and if this would impact the α-factor. Nitrite accumulated in R-2 at DO concentrations ranging from 0.50 to 7.35 mg L^{-1} and at free ammonia (FA) concentrations ranging from 1.34 to 7.19 mg L^{-1}. Nonsteady-state re-aeration tests performed on the effluent from each reactor and on tap water indicated that the α-factor increased as SRT increased. A simple statistical analysis (paired t-test) between AOURs and COURs indicated that there was a statistically significant difference at 0.05 level of significance for both reactors. *Mines concluded that the ex situ BOD bottle method for estimating AOUR appears to be invalid in bioreactors operated at low DO concentrations (<1.0 mg L^{-1}).*

7.7.4   Alternate Methods to be Considered

In enzymes technology, bacterial oxidation is hugely related to enzymes and delta G (ΔG), where G is the Gibbs free energy, in relation to their different growth phases, in particular the log-growth phase and the endogenous phase. Since bacteria cells are extremely good model systems, as such, the energy balance of the cells may be used to calculate the oxygen demand, together with techniques such as FISH, 16S RNA and so forth, to estimate the cell numbers.

This will then be an alternative method of estimating the OUR (Oxygen Uptake Rate) in a treatment plant. The OUR is an overriding important parameter that governs the treatment performance. However, even though the idea is laboratory proven and reproducible, in the field, in open vessels with hundreds if not thousands of types of bacteria, with hundreds if not thousands of substrates in wastewater, we do not have the tools to determine OUR in the way proposed. Therefore, the proposer is trying to break through the traditional "black box" mentality by finding an "out of the box" way to estimate oxygen transfer, and he speculated that perhaps Specific Oxygen Uptake Rate (SOUR) would be a tool. Under the right conditions, one may get a very high correlation coefficient between substrate consumption, theoretical oxygen demand, and oxygen transfer, provided that the microbial community is definable in terms of oxygen respiration rates.

However, when looking at degradation potential for a given substrate, one generally looks at both makeup and structure to understand the availability of the carbon and nitrogen to the organisms–single bonds being more available than double bonds; double bonds being more available than triple bonds; and so on. The number of chlorine molecules clustered around a carbon molecule and the number of carbon atoms in a ring are a strong influence on kinetics and on whether the degradation reaction will go; and so forth. All of these things influence the availability of the molecule. As many researchers point out, if we cannot control the substrate, the correlation between uptake rate and oxygen transfer becomes more difficult to sort out. This becomes a research activity rather than an oxygen transfer estimation technique.

However, one might be able to make this work in the same way we do a clean water test by dictating the substrate and the other physical parameters, even though one could not make this work in a generic wastewater. Instead of treating the ecosystem as a black box, it is proposed to look at the microorganisms at the molecular level. It is proposed that this work is to be coordinated by the ASCE/EWRI Oxygen Transfer Standards Committee (OTSC) in conjunction with the USEPA in an effort to upgrade relevant guidelines and standards. By considering the ADP-ATP cellular-energy transfer system, the heterotrophic bacterial metabolism, chemosynthetic autotrophic bacterial metabolism, photosynthetic autotrophic bacterial metabolism, and so forth, it may be possible to play the zero-sum game in the matter of oxygen supply and demand. This may then allow development of a reference standard method for measurement against which all other methods can be judged.

This would then allow sensible methods to be devised and calibrated against the reference standard method. If a baseline mass transfer coefficient can be established, the mass transfer can be calculated at a *baseline* scale. Similarly, by knowing the SOUR, using a reference substrate and a reference microbial make-up, it might be possible to identify valid procedures and methodologies to correlate OTR with OUR. However, at this stage the method is primarily used to validate Eq. (7.1) that states that the oxygen transfer rate (OTR) is indeed related to the respiration rate if such a relationship is true, and is a valid method to confirm the other methods such as the dilution method explained below.

7.7.5 Dilution Method

In the article by Doyle and Boyle (1981), an interesting method of testing for alpha was suggested. It appears that it may be possible to use the dilution method as suggested by the article to test out the determination. By first aerating a tank of pure water (as shown in Fig. 7.5 below) to an elevated DO, say to 7 p.p.m., and then, upon stopping the aeration, gradually and carefully pouring a sample of known volume of the activated sludge mixed liquor into the tank, and then gently mixing them together, it may be possible to measure the slope of the DO decline curve at quiescent conditions, thereby eliminating the first possible explanation for the cause of increased OUR measurement. If the sample has been diluted to 50% by the tank, the resultant slope should then be multiplied by 2 to get the true OUR. This should then be compared with the steady-state column test with an *in-situ* measurement as recommended by the ASCE Guidelines (ASCE 1997). Alternatively, mixed liquor can be continuously pumped to the test tank from a position within the existing aeration basin using a displacement pump until a set known volume is withdrawn.

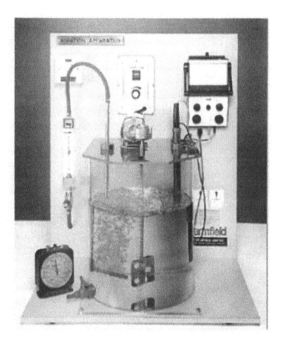

Figure 7.5 A typical bench-scale aeration unit (image from Armfield Ltd.)

This should give the same oxygen depletion curve as the steady-state test, allowing the measurement of the slope of the curve as a measurement of the microbial oxygen uptake rate.

To avoid any substrate limiting effect, the test should be done *in-situ* as quickly as possible just like the off-gas column test. A schematic of the method and apparatus is depicted in the diagrams below (Fig. 7.6a, Fig. 7.6b, Fig. 7.6c).

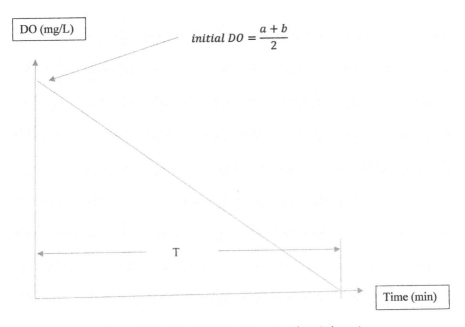

DO (mg/L)

$$initial\ DO = \frac{a+b}{2}$$

T

Time (min)

FIGURE 7.6a Plot of DO (mg/L) versus time (minutes)

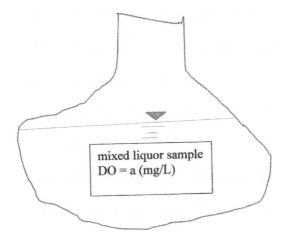

mixed liquor sample
DO = a (mg/L)

FIGURE 7.6b Sample of mixed liquor of a known volume in a container

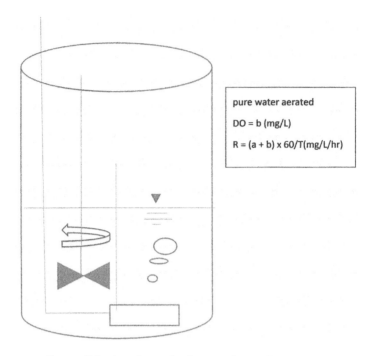

FIGURE **7.6c** Aeration tank of same volume of pure water

7.7.5.1 Steps for the Dilution Method

1. Aerate a tank of pure water of 4 to 6L volume (as shown in Fig. 7.6) to an elevated DO, say to 7 p.p.m.
2. Collect a sample of mixed liquorto be evaluated into a container of approximately 4 to 6 L.
3. Upon stopping the aeration, gradually and carefully pourthe sample of known volume of the activated sludge mixed liquor into the tank.
4. Then gently mix them together by a mechanical stirrer.
5. Immediately measure the DO by the Winkler Method or by using a calibrated fast-response DO probe with probe time constants less than $0.02/K_L a$.
6. Plot the DO versus time on a graph (as shown in Fig. 7.6a) and calculate R using a linear least-squares regression to fit a straight line through the data points. Since the mixed liquor is diluted to 50% its original concentration, the resultant slope of the line is multiplied by 2 to obtain the *in-situ* R value. The time lapse between sample collection and uptake rate measurement is critical in this *ex-situ* procedure. The entire process from collection of sample to starting DO monitoring should take less than 2 min [ASCE 18-18]. The procedure should be replicated at least three times at any given sampling point. A worked example is given below.

Suppose the treatment plant *in-situ* aeration tank has a DO level of 2 mg/L and the bench-scale aeration unit is aerated to 7 mg/L, then the mixture will have a DO concentration of:

$$(a + b)/2 = (2 + 7)/2 = 4.5 \text{ mg/L}$$

Suppose the DO level dropped to zero in 7 mins, then the slope of the decline curve would be:

$$4.5/7 \times 60 = 38.57 \text{ mg/L/hr}$$

The actual oxygen uptake rate *in-situ* is therefore given by twice this value or

$$2 \times 38.57 = 77 \text{ mg/L/hr}$$

This experiment should then be compared with the BOD bottle method to confirm whether the disturbance due to shaking was the real cause of over-estimation or that it was caused by increased oxygen availability.

7.7.6 Procedure for Determination of Wastewater Mass Transfer Coefficient $K_L a_f$ and Alpha (α)

The following procedure can be used in the bench-scale determination of the oxygen transfer coefficient, α:

1. A vessel similar to those shown in Fig. 7.7 can be used. For the diffused system, where the gas-side gas depletion effect is significant, an air measuring rotameter can be installed for recording the gas flow rate.
2. The container is filled with a defined volume of tap water or wastewater as the case may be, and the water temperature and air barometric pressure are recorded. The water is then deoxygenated as per ASCE standard 2-06, either chemically or physically stripping the oxygen from solution with an inert gas such as nitrogen. In the laboratory unit, the latter is preferred in order to eliminate any possible chemical interferences.
3. Once the contents have been deoxygenated, the water is re-aerated at a controlled diffused air flow together with any mechanical rotational speed if used to increase the transfer rate. If the latter is employed, care must be taken to ensure that the device simulates the actual device used in the field (this would require separate modeling). The re-aeration curve can be plotted and the important parameters $K_L a$ and C_∞ can then be estimated using the non-linear least squares (NLLS) method as described in the standard. This step can be repeated for various gas flow rates, mixing intensities, temperatures and pressures.
4. If wastewater is used in the test, the procedure should come after the clean water test, and should be repeated using the same volume of wastewater as the previous cleanwater. As the test results would depend on the microbial respiration rate, as previously described, every effort should be made to eliminate the influence of the microbial respiration.

FIGURE 7.7 Laboratory apparatus for determination of α [Eckenfelder 1970]

This can be achieved by killing off all live microbes, such as using a sulphamic acid-copper sulphate solution, or if a bench scale biological reactor has been used, the effluent from the reactor should be used for the test when a completely mixed system is contemplated either in the entire aeration basin, or a section of the basin where completely-mixed is envisaged. If the latter method is used, the biological solids should be filtered out to prevent any microbial oxygen uptake that remains during the test period. The respective K_La values as determined from the two tests (clean water vs. wastewater) can then be compared at the same temperature, pressure and mixing conditions, and α can be calculated in accordance with the following equation:

$$\alpha = (K_La \text{ wastewater})/(K_La \text{ clean water}) \qquad (7.18)$$

Should all the methods described above give the same alpha value and the same in-process mass transfer coefficient at full scale, this proposal would be a success and the standards and guidelines for oxygen transfer measurement should be amended accordingly.

7.8   ASSOCIATED COST-BENEFIT IMPLICATIONS

7.8.1   Upgrading of Current Standards

The validity of this proposal is based on two concepts:

1. the oxygen transfer rate in the field under process condition is affected by the microbial oxygen uptake rate for a diffused aeration system;
2. the alpha-factor (α) is only dependent on the wastewater characteristics.

If these concepts are correct, then the oxygen transfer rate calculations must take into account the respiration rate at the time of measurement. This would result in replacing Eq. CG-1 in the clean water measurement standard ASCE 2-06 by Eq. (7.1) recalling as follows:

$$OTR_f = \left(\frac{1}{(C_\infty^*)_{20}}\right)\left[\alpha(SOTR)\theta^{T-20}\right]\left(\tau \cdot \beta \cdot \Omega \cdot (C_\infty^*)_{20} - C\right) - RV$$

where SOTR is determined as per Eq. 7.10 above; C is the DO concentration at any process DO level; R is the respiration rate at this process level. Both the parameters SOTR and the oxygen saturation concentration C_∞^* are based on a series of clean water tests at pilot scale or shop tests as described in the section Technical Approach. It is, however, required to determine the baseline clean water mass transfer coefficient $K_L a_0$ in order to extrapolate this coefficient to full scale. Alpha is determined in bench scale as shown by Fig. 7.7 and by using Eq. 7.18 and Eq. 7.20 below. The mass transfer coefficient at full scale under process condition is estimated by Eq. 7.19 at any determined alpha value in the laboratory at bench scale. The other correction factors are as per the current standard. The cost-benefit implication is that there is no longer any need to determine the essential parameters at full scale, since at the outset of a design project, full scale aeration tanks are not available prior to design. With the current standard, the discrepancies between anticipated and actual performance are often sufficiently large to warrant substantial field modifications to the aeration equipment furnished. The costs of performing such modifications and the ill will generated testify to the need for improved oxygen transfer design and testing procedures. Of the other correction factors, the temperature correction factor θ should be based on the 5th power model as advanced by Lee (2017) and not just using $\theta = 1.024$ as per the standard that has many fallacies.

The analysis in Yunt's experiment [Yunt et al. 1980], based on the FMC diffusers, appears to support the temperature correction model [Lee 2017] as shown by Fig. 7.8 (also Fig. 3.7a), showing the excellent regression correlations when the baseline is used with the model. Using the baseline $K_L a_0$ is tantamount to using a shallow tank, which is the fundamental basis for the 5th power temperature model. However, the temperature model used in conjunction with the depth correction model allows scaling up to a higher water depth tank. In biotechnology, temperature correction is very important, as the standard and current guidelines have pointed out in various sections. Currently, reported values for θ range from 1.008 to 1.047. Because it is a geometric function, large error can result if an incorrect value of θ is used. While the Arrhenius equation is used substantially for the wastewater and other industries, and is applicable to cooler and temperate climates, it should be used with caution at higher temperatures. At temperatures below 20°C, it is not important which equation is used, but in countries with hot climates, and especially those with high humidity, where the effect of evaporative cooling is reduced, the Arrhenius equation may not be applicable and could result in systems that do not perform to design specifications. Furthermore, temperature correction is affected by size of bioreactors, aeration tanks, contaminants, and mixing intensities. Using the proper temperature model would enhance the accuracy of the calculation of the standard specific baseline as demonstrated in Fig. 5.7 for the FMC diffusers, whereas the Arrhenius formula may be less exact, as shown in Fig. 5.8 for the Norton diffusers.

Therefore, this project has the additional benefit of confirming the validity of the 5th power temperature correction model as well as the other models concerning the effect of geometry on the mass transfer coefficient. For the upgrading of the standard, the research should additionally be done under the umbrella of a university or similar institution, and subjected to peer review. If the results are believed to be applicable for a standard, it could then be submitted to the Standards Committee for consideration and further verification.

7.8.2   Translating Clean Water Test Results to In-process Water Measurements

It is postulated that this correction factor (α) for wastewater can be determined by bench scale experiments. Eq. (3.6) stated in Chapter 3 for the main model can be plotted for $K_L a_f$ against the function ϕZ_d for when the baseline is unity, for various α values, as shown in Fig. 7.3 (also Fig. 3.10 and Fig. 6.2). This graph shows exactly what Boon (1979) has found in his experiments that $K_L a_f$ is a declining trend with respect to increasing depth of the immersion vehicle of gas supply. The generic term for the mass transfer coefficient is symbolized as $K_L B$.

where $K_L B = K_L a$ for clean water ($\alpha = 1$) and $K_L B = K_L a_f$ for wastewater ($\alpha < 1$).

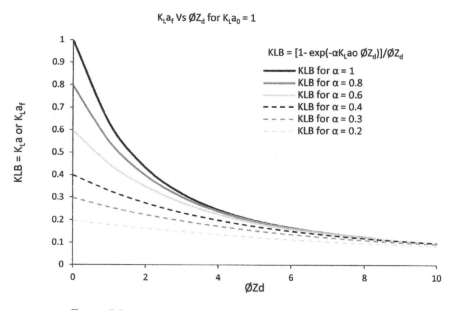

FIGURE 7.8 Apparent mass transfer coefficient vs. function of Z_d

Note that when $\alpha < 1$, and in the presence of microbes, the biological floc exerts a resistance force (F_{bf}) to the gas transfer, so that by the principle of superposition, the net transfer in the field would be given by OTR_f where $OTR_f = SOTR - F_{bf}$. This resistance force is equivalent to the additional gas-side gas depletion rate in the field (gdp_f) and can be measured by the off-gas method applied to an aeration basin. At steady state, this gas depletion is the same as the respiration rate R, and so OTR_f would be given by SOTR calculated by Eq. 7.11 and Eq. 7.12 and re-converting to field conditions, and then subtracting the respiration rate over the entire volume (RV) as shown by Eq. 7.1 [Lee 2019b].

In the application to wastewater treatment, using the transfer of oxygen to clean water as the datum, it may then be possible to determine the equivalent bench-scale oxygen transfer coefficient $(K_L a_{0f})$ for a wastewater system, where the subscript f denotes in-field process water, and the ratio of the two coefficients can then be used as a correction factor to be applied to fluidized systems treating wastewaters via aerobic biological oxidation. It is paramount to determine alpha (α) where alpha is the correction factor (Rosso and Stenstrom 2006b) given by:

$$\alpha = \frac{K_L a_{0f}}{K_L a_0} \approx \frac{K_L a_f}{K_L a} \tag{7.19}$$

As it is hypothesized that this alpha value is not dependent on the liquid depth and geometry of the aeration basin, it is therefore postulated that this correction factor (α) can be determined by bench scale experiments, and the model developed that relates $K_L a$ to depth then allows the α value to be used for any other depths and geometry of the aeration basin. Therefore, after incorporating α into the baseline mass transfer coefficient for clean water, the mass transfer coefficient in in-process water $K_L a_f$ would be given by Eq. 3.6 and repeated here upon incorporating the alpha-factor as:

$$K_L a_f = \frac{1 - \exp(-\Phi Z_d \cdot \alpha K_L a_0)}{\Phi Z_d} \tag{7.20}$$

The mass transfer coefficient so determined should match the result of field testing by following the ASCE Standard Guidelines for In-Process Oxygen Transfer Testing methods [ASCE 1997].

The outcome of this project, if successful, would challenge the current concept that there is absolutely no way to relate alpha to any of these affecting variables other than full scale testing for every condition [ASCE/EWRI 2-06] because of the difficulties in calculating alpha [Mahendraker et al. 2005].

If the testing was done on an old diffuser, then α would be replaced by $\alpha \cdot F$ where F is a fouling factor as defined in the standard.

It should be noted in passing that, in the above equation, the parameter $\emptyset$ depends on Henry's Law constant. Therefore, this equation is equivalent to Eq. 3.6, where $\emptyset$ is $x(1-e)$. If the handbook data of oxygen solubility is plotted against the inverse of the temperature correction function affecting solubility, the straight-line linear plot would be as shown in Figure 7.9 below.

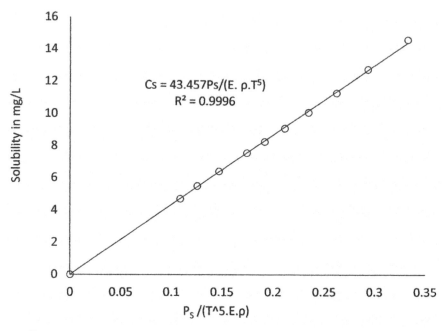

FIGURE 7.9 Solubility plot for water dissolving oxygen at $P_S = 1$ atm (1.013 bar)

Therefore, the solubility law [Lee 2017] can be expressed either by the equation derived from plotting the insolubility or expressed by the equation from plotting the data as in Figure 7.9. In the former method, the equation gives the insolubility of oxygen which is expressed by:

$$\left(\frac{1}{C_S} = 0.02302.T^5 \times E \times \frac{\rho}{P_S}\right) \tag{7.21}$$

where T is in K (Kelvin) to the power 10^{-3}.

In the latter case, the equation gives the solubility directly and is expressed by:

$$C_S = 43.457 \times \frac{P_S}{(T^5 E \rho)} \tag{7.22}$$

Henry's Law is applicable only to ideal solutions [Andrade 2013], and for an imperfect liquid subject to changes in physical state, at extreme temperatures between 273 K and 373 K, it is only approximate and limited to gases of slight solubility in a dilute aqueous solution with any other dissolved solute concentrations not more than 1 percent. Since in Henry's Law, the solubility C_S is proportional to the partial pressure, the Henry's Law constant would be given by $H = 43.457/ (T^5 E\rho)/Y_0$, where $Y_0 = 0.2095$ for air. In a mixed liquor, the liquid density may depart from that of clean water significantly, and so the sample should be measured for its density using a hygrometer to measure the humidity and a hydrometer to measure the relative density.

The discovery of a standard baseline $(K_La_0)_{20}$ that may be determined from shop tests for predicting the $(K_La)_{20}$ value for any other aeration tank depth and gas flowrate, and even for in-process water with an alpha (α) factor incorporated into the equations, is important not only in the development of energy conservation for wastewater treatment plants, but also in the prediction of field in-process performance of an aeration tank in a utility setting. Such prediction would depend on the veracity and validity of Eqs. (7.1, 7.19 and 7.20) on which this proposal is based and on the assumption that the respiration rate R can be accurately measured, as the following worked example illustrates:

Worked Example

The proposed method of calculating the oxygen transfer rate in the field (OTR_f) is best illustrated by a mock example as shown below. Using the test results for the Norton diffusers as shown in Table 5.8, the derived baseline mass transfer coefficient can be obtained to be 0.1076 min^{-1} as shown in Table 7.1 below, for a gas flowrate of 127 scmh (standard cubic meters per hour). This baseline value is used to design an aeration basin of the same surface area but 9 m deep, based on the following hypothetical test conditions:

$V = 37.2 \text{ m}^2 \times 9 \text{ m} = 335 \text{ m}^3$;

C (the mixed liquor DO concentration) = 2 mg/L;

measured average oxygen uptake rate $R = 20$ mg/L;

mixed liquor temperature $T = 20°C$;

barometric pressure $P_a = 101.3$ kPa;

alpha-factor $(\alpha') = 0.6$ derived by testing devoid of living cells as described in section 7.1.5 (see Fig. 7.7). The calculation steps are given below:

(i) Determine K_La_f for the 9 m Tank

In the application of Eq. 7.20 for the full-scale mass transfer coefficients, the first step is to calculate $\emptyset \cdot Z_d$. As given by Eq. 7.22, Henry's Law constant is calculated by $H = 43.457/(T^5 E\rho)$ but since the bioreactor is only lightly fed, the modulus of elasticity and density of the mixed liquor are assumed to be same as pure water, and so it can be read from a chemistry handbook to give $H = 4.382 \times 10^{-4}$ (mg/L)/(N/m^2) compared with 4.383×10^{-4} using the Eq. 7.22. Therefore, the hydrostatic pressure at diffuser depth is given by,

$$P_d = 101325 + 9789 \times 9 - 2333 = 187093 \text{ Pa};$$

and the hydrostatic pressure at equilibrium depth, assuming mid-depth $(e = 0.5)$, is given by:

$$P_e = 101325 + 9789 \times 9/2 - 2333 = 143043 \text{ Pa};$$

the height-averaged gas flowrate is given by Lee (2017) and also given in Chapter 2 Eq. 2.25 as

$$Q_a = Q_S \times 172.82 \times 293.15 \times (1/101325 + 1/187093) = 0.77Q_S$$

where Q_S is given in Table 7.2 as 127 scmh. Therefore, Q_a can be estimated to be 0.77 × 127/60 = 1.63 m³/min.

Therefore,

$$\phi = x(1 - e)$$

where

$$x = HR'ST/Q_a = 4.382\text{E-}4(0.260 \text{ kJ/kg-K}) (37.2 \text{ m}^2) (293.15 \text{ K})/1.63 \text{ m}^3/\text{min}$$
$$= 0.7622 \text{ min/m}$$

and

$$\Phi = 0.7622(1 - 0.5) = 0.3811 \text{ min/m}.$$

This gives $\phi Z_d = 3.43$ min.

Hence, from Eq. 7.20, the mass transfer coefficient in the field for this scaled-up tank would be:

$$K_L a_f = (1 - \exp(-\alpha \cdot K_L a_0 \times \phi Z_d))/\phi Z_d$$

calculated as follows:

$$K_L a_f = (1 - \exp(-0.6 \times 0.1076 \times 3.43))/3.43 = \mathbf{0.0579 \text{ min}^{-1} = 3.48 \text{ hr}^{-1}}.$$

Assuming the equilibrium mole fraction Y_e is about 0.2000, $C^*_{\infty f} = HY_e P_e = 4.382\text{E-}4 (0.2000) (143043) = 12.54$ mg/L (this value should be corrected for β as per ASCE 2007. Assuming $\beta = 0.966$, this gives $C^*_{\infty f} = 0.966 \times 12.54 = 12.12$ mg/L). This saturation concentration will be used to calculate the OTR$_f$ as a first approximation in Eq. 7.15. Another method is to use the spreadsheet model to calculate the Y_e for the clean water data at saturation, as shown in step (iv) below and using the same β factor for $C^*_{\infty f}$.

(ii) Determine $K_L a$ for the 9 m Tank

The clean water mass transfer coefficient for this 9 m tank can be calculated by Eq. 7.20 again with $\alpha = 1$ as follows:

$$K_L a = (1 - \exp(-K_L a_0 \times \phi Z_d))/\phi Z_d.$$

Hence $K_L a = (1 - \exp(-0.1076 \times 3.43))/3.43 = 0.08998 \sim \mathbf{0.0900 \text{ min}^{-1} = 5.40 \text{ hr}^{-1}}$.

(iii) Determine α'

Therefore, the calculated alpha value (α') would be given by $\alpha' = 0.0579/0.0900 = 0.64$ which is quite close to the experimental value of 0.6, thus verifying that Eq. 7.19 is practically valid. The above calculations also prove that the hypothesis $\alpha' = 2\alpha$ is correct, since $\alpha = 0.34$ (see step (iv) below) and $\alpha' = 2 \times 0.34 = 0.68$ which is close to the α' of 0.64 as calculated by the proposed model.

(iv) Determine OTE_f and α

Therefore, from Eq. 7.15, written as

$$OTR_f = K_L a_f (C_{\infty f}^* - C)V - RV$$

$OTR_f = [3.48\,(12.12 - 2)\,335 - 20 \times 335] \times 10^{-3} = 5.097$ kg/hr at a mass gas flowrate of 127 scmh. The OTE_f is given by $5.097/(127 \times 1.20 \times 0.23) = 0.1457$ or 14.57%. This efficiency should correspond to the field test result such as using the off-gas method. From Table 5.8, the Norton clean water standard OTE was calculated to be 42.85% for the 6 m test tank; therefore, the ratio of efficiencies would be alpha (α) = $14.57/42.85 = 0.34$ approximately if the cleanwater efficiency holds for the 9-m tank as well. A more precise method is to use the spreadsheet to simulate the parameters by solving all the model equations for the scale-up tank (Eqs. 3.6 ~ 3.10) as shown in the following calculation sheet (Table 7.2) below, where the estimated result for the clean water mass transfer coefficient is $K_L a = 5.33$ hr^{-1} and the equilibrium DO concentration would be $C_\infty^* = 11.59$ mg/L. Therefore, the clean water OTR at 2 mg/L would be given as $5.33\,(11.59 - 2) \times 335 \times 10^{-3} = 17.1$ kg/hr. Since $W_{O2} = 127 \times 1.20 \times 0.23 = 35.05$ kg/hr [ASCE 2007], the OTE = $17.1/35.05 \times 100 = 48.8\%$ which is higher than the previously assumed value of 42.85%. Since $\beta = 0.97$, $C_{\infty f}^*$ would become $0.97 \times 11.59 = 11.24$ mg/L.

Therefore, from Eq. 7.15, $OTR_f = K_L a_f (C_{\infty f}^* - C)V - RV = [3.48\,(11.24 - 2)\,335 - 20 \times 335] \times 10^{-3} = 4.072$ kg/hr at a mass gas flowrate of 127 scmh. Then OTE_f becomes $4.072/35.05 = 11.6\%$. Therefore, alpha would be given by: $\alpha = 11.6/48.8 = 0.24.$

(v) Determine the Respiration Rate R and Compare with Measured OUR

Comparing this estimated α with the model result that gives $\alpha' = 3.48/5.33 = 0.65$ as determined in step (iii), which, although quite close to the measured α' of 0.60, is at least more than twice the value of alpha (α) of 0.24. If it is desired to balance the two estimations assuming $\alpha' = 2\alpha$ (for the steady state), then the measured OUR_f should be around 16.8 mg/L/hr instead of 20 mg/L/hr.

In a previous paper, the proposer derived an equation that links up the two parameters alpha (α) and alpha' (α') at any value of DO concentration as follows [Lee 2019b] and also given by Eq. 6.28 in Chapter 6:

$$\alpha' = \alpha + \frac{R}{K_L a\,(\text{deficit})}$$

Rearranging this equation gives: $R = (\alpha' - \alpha)\,K_L a\,(\beta C_\infty^* - C)$

Therefore, $R = (0.65 - 0.24)\,5.33\,(11.24 - 2) = 20$ mg/L/hr which is greater than 16.8 mg/L, indicating that the test was not done at steady state.

TABLE 7.1 Example calculation for the baseline K_La_0 based on Norton diffusers data (Eqs. I to V are identical to Eqs. 3.6 to 3.10)

Test Date	5/16/1978	Norton diffuser						
Environmental Data								
	Water temp.	$T =$	20	degC				
	atm press.	$P_a =$	101325	N/m^2				
	Tank area	$S =$	37.2	m^2		**Error Analysis**		
Calc. Variables						Eq. I =	−2.21E-04	4.89E-08
	Diff press.	$P_d =$	152696	N/m^2		Eq. II =	−9.97E-04	9.94E-07
	$x = HRST/Q_a$	$x =$	0.7182	min/m		Eq. III =	−5.09E-05	2.59E-09
Solver Soln.	Baseline K_La	$K_La_0 =$	**0.1076**	1/min(hr)	6.45	Eq. IV =	1.13E-04	1.27E-08
		$n =$	**2.77**	–		SS (sum of squares) =		1.05832E-06
		$m =$	**1.63**	–		offgas		
Equil. mole fraction		$y_e =$	**0.2067**	–		Eq.V =	0.2095	checked
Diff mole fraction		$y_d =$	0.2095	–				
	Eff. depth	$d_e =$	2.79	m				
	Depth ratio	$e =$	0.51	–				
	Press. at de	$P_e =$	126324	N/m^2				
Input Data		Z_d (m)	Q_a (m^3/min)	K_La (1/min)	C_{inf}^* (mg/L)	r_w (N/m^3)	H (mg/L/(N/m^2))	p_{vt} (N/m^2)
		5.49	1.73	0.0970	11.44	9789	4.382E-04	2333

TABLE 7.2 Calculation sheet for 9 m tank using Norton diffuser at 1.63 m^3/min (127 scmh) for clean water aeration at $P_b = 1$ atm (1.013 bar) and standard temperature of 20°C

Fixed							SS error
T (deg C)	$Z_d =$	9.00	Q_a (m^3/min)	1.63	$de =$	4.0979	
20	$P_a =$	101325	$T(K)$	293.15	$Pe =$	139106	
	$P_d =$	187093	$Q_S =$	127 scmh	Eq. 1	5.9024E-04	3.48389E-07
	$x =$	0.7624			Eq. 2	−3.3022E-06	1.09046E-11
	$S =$	37.2			Eq. 3	1.8525E-05	3.43165E-10
Variables					Eq. 4	−3.6811E-04	1.35503E-07
	$K_La =$	**0.0890**	5.33 hr^{-1}		Eq. 5	2.1676E-08	4.69838E-16
	$e =$	**0.46**			Eq. 6	3.0261E-06	9.15723E-12
	$n =$	**2.33**					4.84255E-07
	$m =$	**1.47**	Eq. I, II, III, IV, V, VI given by:				
	$Ye =$	**0.1901**	(1) $K_La = (1 - \exp(-mx K_La_0 Z_d))/n/m/x/Z_d + (n-1)K_La_0/n$				
	C_{inf}^*	**11.59**	(2) $C_\infty^* = nH*Y_d*(P_a - P_{vt} - P_d \exp(-mx K_La_0 Z_d))/(1-\exp(-mx K_La_0 Z_d))$				
Data	$K_La_0 =$	0.1076					
	$Yd =$	0.2095	(3) $\ln(P_e mx K_La_0/n/r_w) + \ln(nHY_d P_d/C_\infty^* - 1) = mx K_La_0 Z_e$				
	$H =$	4.383E-04	(4) $K_La = (1 - \exp(-x(1-e)Z_d K_La_0))/(x(1-e)Z_d)$				
	$r_w =$	9789	(5) $y_0 = C_\infty^*/n/H/(p_a - p_{vt}) + ((Y_d \cdot P_d/(P_a - P_{vt}) - C_\infty^*/(n_H(p_a - p_{vt})))$				
	$p_{vt} =$	2333	$\exp(-mx \cdot K_La_0 Z_d)$				
	$P_S =$	98992	(6) $C_{inf}^* = HY_e P_e$				

The only way to verify the above calculations in the example is by means of an actual testing of wastewater with the appropriate composition, and aerating it inside a 9 m column or tank using the same Norton diffuser to observe the resultant $K_L a_f$ and compare it with the clean water test $K_L a$ using the same column. The respiration rate should be independently measured by a suitable method such as the dilution method as described in section 7.1.4, or the off-gas Column Test as given by ASCE 18-18 Appendix D.

7.8.3   Energy Conservation

About 50 to 85 percent of the total energy consumed in a biological wastewater treatment plant is in aeration. The activated sludge process, the most common process, is performed in large aeration basins and an excess factor of safety is used in designing for the air supply to meet sustained peak organic loading and to avoid endogenous situations. This may lead to unsatisfactory treatment performance and even plant failure.

In terms of the issues of environmental and economic significance of the research, the current improper sizing of the aeration system is primarily due to the inability to estimate the mass transfer coefficient ($K_L a$) correctly for different tank depths, among other things, leading to improper blower design and to inappropriate operation. The new findings have the potential to help against the wasteful energy practice in WWTP (Wastewater Treatment Plants) - supplying the air in the excess, which also has adverse impact on the regime of treatment.

The objective of this proposal is to introduce a baseline oxygen mass transfer coefficient ($K_L a_0$), a hypothetical parameter defined as the oxygen transfer rate coefficient at zero depth, and to develop new models relating $K_L a$ to the baseline $K_L a_0$ as a function of temperature, system characteristics (e.g. the gas flow rate, the diffuser depth Z_d), and the oxygen solubility (Cs). Results of this study indicate that a uniform value of $K_L a_0$ that is independent of tank depth can be obtained experimentally. This new mass transfer coefficient, $K_L a_0$ is introduced for the first time in the literature and is defined as the baseline volumetric transfer coefficient to signify a baseline. This baseline $K_L a_0$ has proven to be universal for tanks of any depth when normalized to the same test conditions, including the gas flow rate U_g (commonly known as the superficial velocity when the surface tank area is constant). The baseline $K_L a_0$ can be determined by simple means, such as a clean water test as stipulated in ASCE 2-06. The developed equation relating the apparent volumetric transfer coefficient ($K_L a$) to the baseline ($K_L a_0$) is mainly expressed by Eq. (3.6).

The standard baseline ($K_L a_0)_{20}$, when normalized to the same gas flow rate, is a constant value regardless of tank depth. This baseline value can be expressed as a specific standard baseline when the relationship between ($K_L a_0)_{20}$ and the average volumetric gas flow rate Q_{a20} is known. Therefore, the standard baseline ($K_L a_0)_{20}$ determined from a single test tank is a valuable parameter that can be used to predict the ($K_L a)_{20}$ value for any other tank depth and gas flow rate (or U_g (height-averaged superficial gas velocity)) by using Eqs. (3.6 ~ 3.10) and the other developed equations, provided the tank horizontal cross-sectional area remains constant and uniform as the bubbles rise to the surface. The effective depth 'de' can be determined by solving a set of simultaneous equations using the Excel Solver or similar software but, in the absence of more complete data, 'e' can be assumed to be between 0.4 and 0.5 (Eckenfelder 1970) for conventional aeration.

Therefore, $(K_L a_0)_{20}$ can be used to evaluate the $K_L a$ in a full-size aeration tank (e.g. an oxidation ditch with a closed loop flow condition) without having to measure or estimate numerically the bubble size needed to estimate the $K_L a$ for such simulation. However, the proposed method herewith may require multiple testing under various gas flow rates, and preferably with testing under various water depths as well, so that the model can be verified for a system. Using the baseline, a family of rating curves for $(K_L a)_{20}$ (the standardized $K_L a$ at 20°C) can be constructed for various gas flow rates applied to various tank depths. The new model relating $K_L a$ to the baseline $K_L a_0$ is an exponential function, and $(K_L a_0)_T$ is found to be inversely proportional to the oxygen solubility $(Cs)_T$ in water to a high degree of correlation. Using a pre-determined baseline $K_L a_0$, the new model predicts oxygen transfer coefficients $(K_L a)_{20}$ for any tank depth to within 1 ~ 3% error compared to observed measurements and similarly for the prediction of the standard oxygen transfer efficiency (SOTE%). The discovery of a standard baseline $(K_L a_0)_{20}$ determinable from shop tests is important for predicting the $(K_L a)_{20}$ value for any other aeration tank depth and gas flow rate, and this finding is expected to be utilized in the development of energy optimization strategies for wastewater treatment plants and also to improve the accuracy of contemporary aeration models used for aeration system evaluations. Hopefully, the problem with energy wastage due to inaccurate supply of air is ameliorated and the current energy consumption practice could be improved by applying the models to estimate the mass transfer coefficient $(K_L a)$ correctly for different tank depths at the design stage.

Notation

α	wastewater correction factor, ratio of process water $\alpha \cdot K_L a$ to clean water $K_L a$
$\alpha \cdot F$	ratio of the process water $(\alpha \cdot K_L a)_{20}$ of fouled diffusers to the clean water $(K_L a)_{20}$ of clean diffusers at equivalent conditions (i.e. diffuser airflow, temperature, diffuser density, geometry, mixing, etc.), and assuming $(\alpha \cdot K_L a)_{20}/(K_L a)_{20} \approx (\alpha \cdot K_L a)/(K_L a)$
α'	wastewater correction factor, ratio of process water (with no microbes) $K_L a_f$ to clean water $K_L a$
β	correction factor for salinity and dissolved solids, ratio of C_∞^* in wastewater to tap water
C_∞^*	oxygen saturation concentration (clean water) in an aeration tank (mg/L)
$C_{\infty f}^*$	oxygen saturation concentration in an aeration tank under field conditions and $R = 0$ (mg/L)
$(C_\infty^*)_{20}$	oxygen saturation concentration (clean water) in an aeration tank at 20°C (mg/L)
$C_{\infty 0}^*$	hypothetical oxygen saturation concentration (clean water) in an aeration tank at zero gas depletion rate (mg/L)

C_S oxygen saturation concentration (clean water) in an infinitesimally shallow tank, also known as oxygen solubility in clean water at test temperature and barometric pressure (mg/L)

C dissolved oxygen concentration in a fully mixed aeration tank (mg/L)

C^* a broad term for any saturation concentration in equilibrium with the partial pressure of oxygen in the gas phase, microscopically or macroscopically

C_R the "apparent" or steady-state saturation concentration in the field (mg/L)

E Modulus of elasticity of water (kN/m^2 × 10^6)

F_{bf} resistance force by the biological floc against oxygen transfer (mg/L/hr)

DO   Dissolved oxygen or dissolved oxygen concentration (mg/L)

k_L mass transfer coefficient in the liquid film (m/min)

k_G mass transfer coefficient in the gas film (m/min)

K_L overall mass transfer coefficient based on the liquid phase (m/min)

$K_L B$ generic term for the apparent mass transfer coefficient measured either for clean water or wastewater (min^{-1} or hr^{-1})

$K_L a$ volumetric mass transfer coefficient (clean water), also known as the apparent volumetric mass transfer coefficient (min^{-1} or hr^{-1})

$(K_L a)_{20}$ volumetric mass transfer coefficient (clean water), also known as the apparent volumetric mass transfer coefficient at standard conditions (min^{-1} or hr^{-1})

$K_L a_f$ mass transfer coefficient as measured in the field, equals $\alpha' \cdot K_L a$ (min^{-1} or hr^{-1})

$K_L a_0$ baseline mass transfer coefficient (clean water), equivalent to that in an infinitesimally shallow tank with no gas side oxygen depletion (min^{-1} or hr^{-1})

$(K_L a_0)_{20}$ baseline mass transfer coefficient (clean water) at standard conditions, equivalent to that in an infinitesimally shallow tank with no gas side oxygen depletion at standard conditions (min^{-1} or hr^{-1})

$K_L a_{0f}$ wastewater baseline mass transfer coefficient, equivalent to that in an infinitesimally shallow tank without gas side oxygen depletion (min^{-1} or hr^{-1})

$(K_L a_{0f})_{20}$ wastewater baseline mass transfer coefficient at standard conditions, equivalent to that in an infinitesimally shallow tank with no gas side oxygen depletion (min^{-1} or hr^{-1})

$\theta \cdot \tau \cdot \beta \cdot \Omega$ temperature, solubility, pressure correction factors as defined in ASCE 2007

MLSS   mixed liquor suspended solids (mg/L or g/L)

OTE   oxygen transfer efficiency (%)

p. SOTE   predicted standard oxygen transfer efficiency (%)

rpt. SOTE   reported standard oxygen transfer efficiency (%)

OTR   oxygen transfer rate (kg O$_2$/hr)

OTR$_f$ oxygen transfer rate in the field under process conditions, which equals: $K_L a_f (C^*_{\infty f} - C)$ V-RV (for batch process) (kg O$_2$/hr)

OTR_{cw}	oxygen transfer rate under non-steady state oxygenation in clean water (kg O_2/hr)
OTR_{ww}	oxygen transfer rate under non-steady state oxygenation in wastewater without any microbial cell respiration (kg O_2/hr)
SOTR	standard oxygen transfer rate in clean water/wastewater as (Fig. 7.3 and Eq. 7.12) also defined in ASCE 2007 standard for clean water (kg O_2/hr)
OUR	oxygen uptake rate (kg O_2/hr)
OUR_f	oxygen uptake rate in the field, measurable by the off-gas method (kg O_2/hr)
R	microbial respiration rate, also known as the oxygen uptake rate (OUR) (kg O_2/m^3/hr usually expressed as mg/L/hr)
R_0	specific gas constant for oxygen (kJ/kg-K)
ρ	density of water (kg/m^3)
σ	surface tension of water (N/m)
z	depth of water at any point in the tank measured from bottom, see Fig. 7.1 (m)
Z_d	submergence depth of the diffuser plant in an aeration tank (m)
Z_e	equilibrium depth at saturation measured from bottom, see Fig. 7.1 (m)
P_S	*net* surface atmospheric pressure, $P_a - P_{vt}$ (kPa)
P_a, P_b	atmospheric pressure or barometric pressure at time of testing (kPa)
P_e	equilibrium pressure of the bulk liquid of an aeration tank (kPa) defined such that: $P_e = P_a + r_w d_e - P_{vt}$ where the following symbols definition apply for $P_{vt} d_e r_w$
P_{vt}	the vapor pressure
r_w	specific weight of water in kN/m^3 (at test temperature)
d_e	effective saturation depth at infinite time (m)
Y_e	oxygen mole fraction at the effective saturation depth at infinite time
Y_0, Y_d	initial oxygen mole fraction at diffuser depth, Z_d, also equal to exit gas mole fraction at saturation of the bulk liquid in the aeration tank, $Y_0 = 0.2095$ for air aeration
H	Henry's Law constant (mg/L/kPa) defined such that:

$$C_\infty^* = HY_e P_e \quad \text{or} \quad C_S = HY_0(P_S)$$

Y_{ex}	exit gas or the off-gas oxygen mole fraction at any time
y	oxygen mole fraction at any time and space in an aeration tank defined by an oxygen mole fraction variation curve
gdp	gas-side oxygen depletion rate (equals OTR; equals zero at steady state for clean water; equals OTR_f in process water; equals gdp_f at steady state) (kg O_2/hr)
gdp_{cw}	gas-side oxygen depletion rate in clean water (equals OTR_{cw}) (kg O_2/hr)
gdp_{ww}	gas-side oxygen depletion rate in wastewater (equals OTR_{ww}) (kg O_2/hr)
gdp_f	specific gas-side oxygen depletion rate due to microbes-induced resistance (equals OTR_f at steady state); equals the microbial respiration rate R in process water (kg O_2/hr)
Q_a	height-averaged volumetric air flow rate (m^3/min or m^3/hr)
S	horizontal cross-sectional area of an aeration basin or tank (m^2)

T	absolute temperature in Kelvin (K) or in °C
U_g	superficial gas velocity (m/min)
V	volume of aeration tank (m³)
Q_S, AFR	gas (air) flow rate at standard conditions (20°C for US practice and 0°C for European practice), in (std ft³/min or Nm³/h for scmh)
n, m	calibration factors for the Lee-Baillod model equation for the oxygen mole fraction variation curve
q	exponent in the mass transfer coefficient vs. gas flow rate curve
W_{O2}	mass flow of oxygen in air stream (kg/h)

7.9 APPENDIX

The differentiation between the mass transfer coefficient and the baseline has been described in Chapter 3, under the section 'Background'. Here the author gives a further elaboration on one of the many myths that has hindered, for many years, the development of new techniques in the clean water measurement and dirty water applications of this technology.

7.9.1 The Myth of 'True' $K_L a$

In the 70s and 80s, it had been correctly recognized that the saturation concentration C^* is a function of both space and time for bubble aeration [Baillod 1979] [Boon and Lister 1973] [Downing and Boon 1968]. What was neglected is that the mass transfer coefficient $K_L a$ is also a function of both space and time since $K_L a$ and C^* are the two sides of the same coin [Lee 2017, 2018]. Therefore, in the standard model, the accumulation term would be given by:

$$dC/dt = \hat{K_L a}(\hat{C^*} - \hat{C}) \qquad (A1)$$

where both $K_L a$ and C^* are variables (the symbol ^ indicates a point in space and time), even when C is uniform at any point in time in a well-mixed tank. By treating $K_L a$ as a constant with respect to space and time however, the following equation is obtained:

$$dC/dt = K_L a(\overline{C^*} - C) \qquad (A2)$$

where the bar indicates an average value over the height of the bulk volume. Eq. A2 was then equated with the standard model applied to a bulk liquid for a bulk aeration system that is described by the terms (bulk-averaged saturation concentration C^*_∞) and (the apparent bulk mass transfer coefficient $K_L a'$), both of which can be obtained by a clean water test, giving the following equality equation:

$$dC/dt = K_L a'(C^*_\infty - C) \qquad (A3)$$

$K_L a$ is then termed the 'true' mass transfer coefficient, as opposed to the 'apparent' mass transfer coefficient $K_L a'$. Conceptually, both coefficients are 'true' $K_L a$ but applied to different scenarios. The former equation (Eq. A2) was deemed to apply to an average saturation concentration over the tank height at a unique point in time during the reaeration test (not a unique point in space), but this equation cannot be valid since the mass transfer coefficient also varies with space and time as the bubble composition changes throughout the test. Apart from the composition, the bubble volume is also a function of time and space, firstly due to the varying oxygen mole fraction, and secondly due to the hydrostatic pressure. Both the liquid film coefficient K_L and the interfacial bubble-surface area 'a' are functions of space and time – any changes in the bubble radius will affect the bubble film thickness and the surface tension, changing the value of K_L, and the interfacial area changes as well during the bubble ascent to the surface. In fact, in a bulk liquid volume, the bubble size distribution within the bulk volume becomes important to the bulk overall mass transfer coefficient. Adjusting a point oxygen saturation concentration C^* to a height-averaged value $\overline{C^*}$ without doing the same for K_La is fundamentally flawed. Treating K_La as a constant is therefore erroneous in Eq. A2, and so *the two equations (Eq. A2 and Eq. A3) cannot be equated to each other unless both parameters are adjusted with respect to height, simultaneously. Hence, the concept of a 'true' K_La is totally unjustified. Similarly, the term 'apparent' K_La makes no sense at all.*

7.9.2   Principle of Superposition and the Concept of a Baseline K_La

"The superposition principle, also known as superposition property, states that, for all linear systems, the net response caused by two or more stimuli is the sum of the responses that would have been caused by each stimulus individually. [Wikepedia]"

One of the major conflicts between the ASCE standard and the European standards is the mathematical treatment of the oxygen saturation concentration in the CWT. This is because the relationship between K_La and C_S is not fully understood by any of the standards. These two functions are in fact the two sides of the same coin. When the water molecules are bonded as a liquid medium, the inter-molecular forces can change according to environmental conditions. When these bonds are weakened, gas can enter the liquid easily and so the rate of gas transfer (as exemplified by the K_La parameter) increases. At the same time, weakened forces between the water molecules means that the water cannot hold as much molecular gas entering the system, and so the solubility of the gas decreases. A simple experiment with a beaker of water would illustrate the fact. If the beaker is initially devoid of dissolved oxygen when subjected to oxygen dissolution by diffusion, the amount of oxygen transfer is the product of K_La and C_S over the beaker volume. This product is a constant no matter how the inter-molecular forces change, as can be demonstrated by changing the temperature (within the normal working range) that would affect the kinetic energy that would change the forces [Eckenfelder 1952] [ASCE 1997] [ASCE 2006] [Vogelaar 2001]. The initial rate of transfer is therefore given by $dC/dt = C_S \cdot K_La$. This is the standard model of oxygen transfer in its most basic form.

Whereas the saturation concentration is related to Henry's law that governs solubility of the gas into liquid, the mass transfer coefficient is not related to Henry's Law. Both parameters, however, are related to the molecular attraction between the water molecules. When the attraction force is strengthened (such as when the temperature is reduced), the capacity to hold more oxygen molecules increases, and so the solubility increases. Conversely, when the attraction force is decreased (such as by an increase in temperature), and the inter-molecular bonding is weakened, oxygen molecules enter at a faster rate, but the capacity to hold them becomes smaller. Since the mass transfer coefficient is related to how fast dissolution occurs, the two parameters are inversely correlated when the liquid is disturbed by an external force. In Chapter 2, a detailed dichotomy on the effect of temperature on these parameters is already given.

Like any physics problem, the principle of superposition is a powerful tool in tackling such problems as oxygen transfer in an aqueous solution where many different forces are at play simultaneously. As the dissolved oxygen content builds up in an aeration basin, this build-up of gas dissolution in the aqueous solution exerts a counter-force for diffusion from the liquid to the gas phases. When the principle of superposition is applied to the system, the net rate of oxygen transfer between the aqueous phase and the gas phase is the vector sum of the two opposing forces, depending on which force is more superior. In mathematical form, therefore, the net transfer is given by

$$dC/dt = \sum_{C}^{C_S}(C \cdot K_L a),$$ (A3a)

where 'C' ranges from zero to Cs. This can be written as $dC/dt = K_L a(C_S - C)$ when we make the assumption that the overall liquid film coefficient $K_L a$ does not change regardless of whether the gas molecules are going into or out of the bubble. This assumption is only correct when the height of the aeration basin is small. In the attempt to delineate the effect of height on the mass transfer coefficient, Eq. A2 makes a lot of sense when considering the $K_L a$ at zero height as a baseline case. The author has now termed $K_L a_0$ as a special case for a tank of infinitesimal height, and, since the height is now zero, the corresponding saturation concentration must be identical to its surface solubility, so the universal equation for this baseline case can be written as:

$$dC/dt = K_L a_0 (C_S - C)$$ (A4)

where C_S is the surface DO saturation concentration that can be read from any chemistry handbook table on solubility (ASCE 2007 considered the table provided by Benson and Krause (1984) as the most accurate). This equation, however, still cannot be equated to the bulk transfer equation as determined by a clean water test because these two cases have different gas-side gas depletion rates. (In theory, Eq. A4 has zero bubble gas depletion). However, for tank aeration with gas depletion, Eq. (A3) can be modified to:

$$\frac{dC}{dt} = K_L a_0 (C^*_{\infty 0} - C) - gdp_{cw}$$ (A5)

where $K_L a_0$ is calculated by re-arranging the depth correction model (Eq. 3.6) from a known value of $K_L a$, the 'apparent' $K_L a$ determined by an experiment, as follows:

$$K_L a_0 = -\frac{\ln(1 - K_L a(\Phi Z_d))}{(\Phi Z_d)}$$ (A6)

The parameter $C^*_{\infty 0}$ is the saturation concentration that would have existed without the gas depletion (note that $C^*_{\infty 0}$ is not C_s), and gdp_{cw} is the overall bubble gas depletion rate during a clean water test. This equation is again based on the *Principle of Superposition* in physics where the mass transfer rate is given by the vector sum of the transfer rate as if gdp (gas depletion rate) does not exist, and the actual gas depletion rate which is a negative quantity. The saturation concentration $C^*_{\infty 0}$ cannot be the same as C_s because the latter is the oxygen solubility under the condition of pressure of 1 atmosphere only, while $C^*_{\infty 0}$ corresponds to the saturation concentration of the bulk liquid under the bulk liquid equilibrium pressure, but deducting the gas depletion (this of course cannot happen, since without gas depletion there can be no oxygen transfer). The hypothetical $C^*_{\infty 0}$ must therefore be greater than C^*_{∞}, which in turn is greater than C_s since the former corresponds to a pressure of P_e (see Fig. 7.1 in the main text), while the latter corresponds to the surfical pressure P_a, the atmospheric pressure or the barometric pressure at the time of testing. This method of reasoning allows solving for the transfer rate from the baseline mass transfer coefficient.

Since $K_L a$ is a function of gas depletion, and since every test tank may have different water depths and different environmental conditions, their gas depletion rates are not the same; hence, they cannot be compared without a baseline [Lee 2018]. Furthermore, by introducing the term gdp_{cw}, the oxygen transfer rate based on the fundamental gas transfer mechanism (the two-film theory) can be separated from the effects of gas depletions on $K_L a$. This gas depletion rate cannot be determined experimentally, since gdp_{cw} varies with time throughout the test. Jiang and Stenstrom (2012) have demonstrated the varying nature of the exit gas during a non-steady state clean water test. Therefore, the only equation that can be used to estimate the parameters is still the basic transfer equation given by Eq. A3:

$$\frac{dC}{dt} = K_L a (C^*_{\infty} - C) \tag{A6a}$$

Note that the use of the 'apparent' term is no longer necessary, since the concept of 'true' $K_L a$ is not acceptable (as explained above). Eqs. A3 and A5 are essentially equivalent to each other but expressed differently ($K_L a$ vs. $K_L a_0$). Therefore, by the same token using the *Principle of Superposition and the Principle of Mathematical induction (i.e. if the phenomenon is true for clean water, it must be true for wastewater)*, for in-process water without any microbes, Eq. A3 would become:

$$\frac{dC}{dt} = K_L a_{0f} (C^*_{\infty 0f} - C) - gdp_{ww} \tag{A7}$$

It is postulated that the biological floc exerts biological-chemical reactions, which produce a counter-flow resistance force equal to the gas depletion rate $gdpf$, giving the following equation:

$$\frac{dC}{dt} = K_L a_{0f} (C^*_{\infty 0f} - C) - gdp_{ww} - gdp_f - R \tag{A8}$$

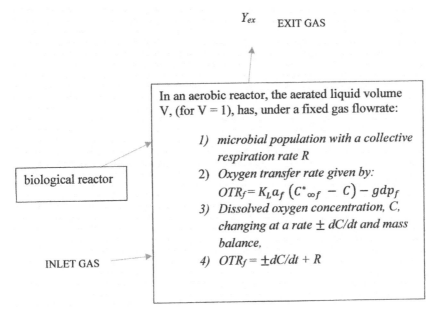

Y_{ex} EXIT GAS

In an aerobic reactor, the aerated liquid volume V, (for $V = 1$), has, under a fixed gas flowrate:

1) *microbial population with a collective respiration rate R*
2) *Oxygen transfer rate given by:*
   $$OTR_f = K_L a_f \left(C^*_{\infty f} - C \right) - gdp_f$$
3) *Dissolved oxygen concentration, C, changing at a rate $\pm dC/dt$ and mass balance,*
4) $OTR_f = \pm dC/dt + R$

biological reactor

INLET GAS

FIGURE A1 Mass balance on the gas phase

Note that the R term in this equation is based on the *Principle of Conservation of Mass* or a material balance, not the Principle of Superposition, as illustrated by Fig. A1 below. [Note: The law of conservation of mass or principle of mass conservation states that for any system closed to all transfers of matter and energy, the mass of the system must remain constant over time, as the system's mass cannot change, so quantity can neither be added nor be removed.] Expressed differently using the measurable parameters, the equation can be written as:

$$\frac{dC}{dt} = K_L a_f (C^*_{\infty f} - C) - gdp_f - R \qquad (A9)$$

where in the above equations, gdp_{ww} is the gas depletion rate for bubbles in wastewater, gdp_f is the gas depletion rate due to the microbial respiration, and R is the microbial respiration rate.

The subscript f refers to field conditions for all the parameters, and that in this last equation, when $dC/dt = 0$, gdp_f would be given by

$$K_L a_f (C^*_{\infty f} - C) - R, \qquad (A9a)$$

where C becomes a constant, usually denoted by a symbol C_R representing the saturation concentration at steady state under process conditions, where Y_{ex} is the off-gas mole fraction that can be measured by an off-gas test [ASCE 1997]. When the system has reached a steady state in the presence of microbes, the gas depletion rate is a constant, and so it would be possible to calculate the microbial gdp by the same equation and by incorporating R as well when $dC/dt = 0$ and $C = C_R$. In the presence of microbes, the advocated hypothesis is that this gdp_f due to the microbes is the same as the reaction rate R and so $dC/dt = K_L a_f (C^*_{\infty f} - C) - R - R$, compared to clean water where the microbial $gdp = 0$. In other words, if F_1 is the gas depletion rate for clean water, and F_2 is the gas depletion rate in process water, then $F_1 - F_2 = R$. It should be noted that, as mentioned before, the basic mass transfer equation is universal, its general form given by the standard model. Therefore, in a non-steady state test for in-process water for a batch test, the transfer equation is given by:

$$\frac{dC}{dt} = K_L a_f (C_R - C) \tag{A10}$$

where C_R is the steady-state DO concentration value attained in the test tank at the *in-situ* oxygen uptake rate, R, under a constant gas supply. But the transfer equation is also given by

$$dC/dt = K_L a_f (C^*_{\infty f} - c) - R - R$$

as mentioned above. Equating the two gives,

$$K_L a_f (C^*_{\infty f} - C) - R - R = K_L a_f (C_R - C) \tag{A11}$$

which gives:

$$K_L a_f = \frac{2R}{(C^*_{\infty f} - C_R)} \tag{A12}$$

Note that in this equation, C is cancelled out, so that the above equation is valid for any value of C, at any state, so long as $dC/dt \geq 0$ and $C < C_R$. Most models do not simulate the gas phase, and so are missing this important element in their balancing equations. This $K_L a_f$ can then be related to the clean water $K_L a$ which serves as a baseline for extrapolating the clean water test results to wastewater by using the correction factors α and β applied to $K_L a$ and C^*_∞, respectively. The correct transfer equation under process conditions at any DO level of C would then be given by Eq. A9 where the gdp_f is given by the microbial respiration rate at a steady state when $C = C_R$, provided that R does not change drastically under another process DO level for the same gas flow rate supplied to the system under test. The estimated value from Eq. A12 (that relies on a measurement of R) for the mass transfer coefficient $K_L a_f$ should then be compared with that determined by a bench-scale or pilot-scale test as depicted in Fig. 7.7 that does not depend on any measurement of R. These two estimations of $K_L a_f$ should give similar results to each other.

REFERENCES

Andrade, Julia. (2013). Solubility Calculations for Hydraulic Gas Compressors. Mirarco Mining Innovation Research Report.

ASCE. (1997). ASCE-18-96. Standard Guidelines for In-Process Oxygen Transfer Testing. ASCE Standard. ISBN-0-7844-0114-4, TD758.S73 1997.

ASCE. (2007). ASCE/EWRI 2-06. Measurement of Oxygen Transfer in Clean Water. ASCE Standard. ISBN-13: 978-0-7844-0848-3, ISBN-10: 0-7844-0848-3, TD458.M42 2007.

ASCE. (2018). ASCE/EWRI 18-18 (2018). American Society of Civil Engineers Standard Guidelines for In-Process Oxygen Transfer Testing. Reston, VA.

Baillod, C.R. (1979). Review of oxygen transfer model refinements and data interpretation. Proc., Workshop toward an Oxygen Transfer Standard, U.S. EPA/600-9-78-021, W.C. Boyle, ed., U.S. EPA, Cincinnati, 17-26.

Benson, B.B., Krause, D. Jr. (1984). The concentration and isotopic fractionation of oxygen dissolved in fresh water and sea-water in equilibrium with the atmosphere. Limnology and Oceanography 29: 620-632.

Boon, A.G., Lister, A.R. (1973). Aeration in deep tanks: an evaluation of a fine bubble diffused-air system. Institute of Sewage Purification 5: 3-18.

Boon, A.G., Lister, A.R. (1975). Formation of sulphide in a rising main sewer and its prevention by injection of oxygen. Progress in Water Technology 7: 289-300.

Boon, A.G. (1979). Oxygen Transfer in the Activated-Sludge Process. Water Research Centre, Stevenage Laboratory, England, United Kingdom.

Chiesa, S.C., Rieth, M.G., Ching, T. (1990). Evaluation of activated sludge oxygen uptake rate test procedures. Journal of Environmental Engineering, Vol. 116, No. 3, May/June, 1990.

DeMoyer, Connie D., Schierholz, Erica L., Gulliver, John S., Wilhelms, Steven C. (2002). Impact of bubble and free surface oxygen transfer on diffused aeration systems. Water Research 37(2003): 1890-1904.

Downing, A.A., Boon, A.G. (1968). Oxygen transfer in the activated sludge process. In: Eckenfelder, W.W. Jr., McCabe, B.J. (eds.). Advances in Biological Waste Treatment. MacMillian Co., NY, p. 131.

Doyle, M., Boyle, W.C. (1981). Translation of Clean to Dirty Water Oxygen Transfer Rates. University of Wisconsin, Madison, Wisconsin (Unpublished).

Doyle, M.L., Rooney, T., Huibregtse, G. (1983). Pilot plant determination of oxygen transfer in fine bubble aeration. Journal (Water Pollution Control Federation) 55(12): 1435-1440. Retrieved from http://www.jstor.org/stable/25042126.

Eckenfelder, W.W. (1952). Aeration efficiency and design: I. Measurement of oxygen transfer efficiency. Sewage and Industrial Wastes, pp. 1221-1228.

Eckenfelder, W.W. (1970). Water Pollution Control. Experimental Procedures for Process Design. The Pemberton Press, Jenkins Publishing Company, Austin and New York. EPA/600/S2-88/022 (1988).

Garcia-Ochoa, F., Gomez, E., Santos, V.E., Merchuk, J.C. (2010). Oxygen uptake rate in microbial processes: an overview. Biochemical Engineering Journal 49(3): 289-307.

Groves, K., Daigger, G., Simpkin, T., Redmon, D., Ewing, L. (1992). Evaluation of oxygen transfer efficiency and alpha-factor on a variety of diffused aeration systems. Water Environment Research 64(5): 691-698. Retrieved from http://www.jstor.org/stable/25044209.

Houck, D.H., Boon, A.G. (1980). Survey and Evaluation of Fine Bubble Dome Diffuser Aeration Equipment, EPA/MERL Grant No. R806990, September, 1980.

Hwang, H.J., Stenstrom, M.K. (1985). Evaluation of fine-bubble alpha factors in near full-scale equipment. Journal WPCF, Volume 57, Number 12, U.S.A.

Keil, Z. Otero, Russell, T.W.F. (1987). Design of commercial-scale gas liquid contactors. AIChE J. 33(3): 488-496.

Lee, J. (2017). Development of a model to determine mass transfer coefficient and oxygen solubility in bioreactors. Heliyon, Volume 3, Issue 2, February 2017, e00248, ISSN 2405-8440. http://doi.org/10.1016/j.heliyon.2017.e00248.

Lee, J. (2018). Development of a model to determine the baseline mass transfer coefficients in aeration tanks. Water Environment Research 90(12): 2126.

Lee, J. (2019a). Baseline mass-transfer coefficient and interpretation of nonsteady state submerged bubble-oxygen transfer data. Journal of Environmental Engineering 2020, 146(1): 04019102.

Lee J. (2019b). Forum: oxygen transfer rate and oxygen uptake rate in subsurface bubble aeration systems. Journal of Environmental Engineering 2020, 146(1): 02519003.

Mahendraker, V., Mavinic, D.S., Rabinowitz, B. (2005). A Simple Method to Estimate the Contribution of Biological Floc and Reactor-Solution to Mass Transfer of Oxygen in Activated Sludge Processes. Wiley Periodicals, Inc. DOI: 10.1002/bit.20515.

Mancy, K.H., Barlage, W.E. (1968). Mechanisms of interference of surface active agents in aeration systems. *In*: Gloyna, E.F., Eckenfelder, W.W. Jr. (eds.). Advances in Water Quality Improvement. Univ of Texas Press.

Mines, R.O., Callier, M.C., Drabek, B.J., Butler, A.J. (2016). Comparison of oxygen transfer parameters and oxygen demands in bioreactors operated at low and high dissolved oxygen levels. Journal of Environmental Science and Health, Part A Toxic/Hazardous Substances and Environmental Engineering 52(4): 341-349.

Rosso, Diego, Stenstrom, Michael K. . Surfactant effects on α-factors in aeration systems. Water Research 40, Elsevier Ltd.

Rosso, Diego, Stenstrom, Michael K. (2006b). Alpha Factors in Full-scale Wastewater Aeration Systems. 2006 Water Environment Foundation.

Wagner, M., Henkel, J., Gunkel, T. (2008). Oxygen Transfer Tests in Clean Water in a Glass Test Tank. Client: Innowater BV/Boxtel, The Netherlands. Certificate.

Yunt, Fred W., Hancuff, Tim O. (1980). Aeration Equipment Evaluation-Phase 1 Clean Water Test Results. Los Angeles County Sanitation Districts, Los Angeles, California 90607. Municipal Environmental Research Laboratory Office of Research and Development. USEPA, Cincinnati, Ohio 45268.

Yunt, F. (1988). Project Summary – Aeration Equipment Evaluation: Phase I – Clean Water Test Results, Water Engineering Research Laboratory Cincinnati OH 45268.

Zhou, Xiaohong, Yuanyuan, Wu, Hanchang, Shi, Yanqing, Song. (2012). Evaluation of oxygen transfer parameters of fine-bubble aeration system in plug flow aeration tank of wastewater treatment plant. Journal of Environmental Sciences 25(2): 295-301. DOI: 10.1016/s1001-0742(12)60062-x. PMID: 23596949.

8

Epilogue

The important findings that have been described in this book are summarized in this chapter as listed below.

8.1 THE STANDARD MODEL

For many years, attempts have been made to develop correlations between $K_L a$ values in particular wastewater and $K_L a$ values in pure water, using the Standard Model for bubble-oxygen transfer. The standard model is given in Eq. 4.1 in Chapter 4 as:

$$\frac{dC}{dt} = K_L a(C_\infty^* - C)$$

Even though the standard model has been in existence since 1924 [Lewis and Whitman 1924], the fundamental principles of the model *applied to a bulk liquid under aeration* have not been fully understood. By definition, the term $K_L a$ must necessarily envelop all of the aerator characteristics associated with oxygenation in a water basin. Hence, the characteristic bubble size, relative velocity, retention time and convective flow patterns are all lumped into this single mass transfer parameter. By applying the *principles of superposition and mathematical induction*, this book has shown that it is possible to mathematically and theoretically derive the equation for the standard model from first principles. The various findings regarding the standard model *applied to a bulk liquid* are summarized below:

8.1.1 Effect of Height of Liquid Column

As a result of this rigorous analysis effort, a mathematical model to determine a *baseline mass transfer coefficient* $(K_L a_0)$ that is independent of the liquid depth, so that all the baseline values determined with any tank height are equal, has been advanced. This method does not require any measurement of bubble size, but does require a typical clean water testing using the ASCE/EWRI 2-06 standard method (ASCE 2007). Conceptually, the baseline coefficient is that which occurs when the tank depth tends to be zero (i.e. where the tank height is virtually non-existent or very small). The mathematical definition of this baseline transfer coefficient is given by:

$$\lim_{Z_d \to 0} K_L a = K_L a_0 \tag{8.1}$$

where Z_d = diffuser depth; $K_L a$ = apparent oxygen mass transfer coefficient; $K_L a_0$ = baseline oxygen mass transfer coefficient. The baseline mass transfer coefficient

$(K_L a_0)$ is defined as the *hypothetical* mass transfer coefficient when the measured $K_L a$ in a typical clean water test is converted to that of an aeration tank of *infinitesimal* height, so that the diffuser immersion depth Z_d tends to be zero. This physical meaning is equivalent to having no gas depletion (*noting that in a gas bubble, Dalton's law states that the total pressure of a mixture of ideal gases is equal to the sum of the partial pressures of the constituent gases. A corollary of this law states that the partial pressure is given by the product of the mole fraction and the total pressure. Under this law, the result of both gradual decreases in hydrostatic pressure and the mole fraction of oxygen as the bubble rises is a gradual decrease in oxygen partial pressure. The latter effect, manifested by the difference between the oxygen content of the feed and exit gases, is termed "exit gas depletion" or simply "gas depletion"*) during gas transfer, similar to surface aeration. The depth correction model, as given by Eq. (4.51), is developed based on mass balances in both the liquid phase and the gas phase as given in the Chapter 4, together with the derivation of the mathematical intricacies.

The effect of changing depth on the transfer rate coefficient $K_L a$ has been explored in this book. As shown in Chapter 4 (Eqs. 4.53 and 4.54), the new model relating $K_L a$ for any tank depth to the baseline $K_L a_0$ is expressed by:

$$K_L a = \frac{1 - \exp(-\Phi Z_d \cdot K_L a_0)}{\Phi Z_d} \qquad (8.2)$$

where Φ is a constant dependent on the aeration system characteristics. This equation is of the same form as Eq. (6.17) given in Chapter 6 for wastewater, and is similar to Eq. 4.44 in Chapter 4 for clean water. By expanding the exponential function into a series, $K_L a = K_L a_0 - K_L a_0^2 \Phi Z_d / 2 + K_L a_0^3 (\Phi Z_d)^3 / 3! \dots$, it can be seen that when Z_d tends to zero, $K_L a$ tends to $K_L a_0$.

The real example given in Chapter 3 (data shown in Table 3.2) [Yunt et al. 1988a], using tanks of 4 different heights 3.05 m (10 ft), 4.57 m (15 ft), 6.09 m (20 ft) and 7.62 m (25 ft) illustrates the validity of the developed model. All the tank heights with the same average volumetric gas flow rate yield an *identical* baseline mass transfer coefficient standardized to 20°C. This baseline coefficient (symbolized as $K_L a_0$) is conveniently expressed as $(K_L a_0)_{20}$ where the subscripts refer to baseline and the standard temperature of 20°C, respectively.

8.1.2   Effect of Gas Solubility

Furthermore, it was found that for any tank depth and temperature (subscript T refers), the mass transfer coefficient, $(K_L a)_T$, is inversely related to the equilibrium saturation concentration, $(C^*_\infty)_T$, with $R^2 = 0.9859$ (Fig. 3.4), in the same way that the baseline $(K_L a_0)_T$ is inversely proportional to the oxygen solubility $(C_S)_T$ (Fig. 3.5, $R^2 = 0.9924$) in water that has already been envisaged in Chapter 3 and in a previous manuscript (Lee 2017). The baseline $K_L a_0$ can be determined by re-arranging Eq. (8.2) to obtain the following:

$$K_L a_0 = -\frac{\ln(1 - K_L a (\Phi Z_d))}{(\Phi Z_d)} \qquad (8.3)$$

The solution for the baseline involves solving Eq. (8.3), together with a set of simultaneous equations as given by Eqs. 4.63, 4.65, 4.68, 4.72 and 4.74. Note that Eq. (4.72) is similar to Eq. (4.66) but is modified to replace y_0 by y_e in Eq. (4.66) to become the following equation to be used in the spreadsheet:

$$C_\infty^* = HY_e(\rho w g \cdot eZ_d + P_b - P_{vt}) \tag{8.4}$$

where Y_e is the mole fraction at the equilibrium level (Fig. 3.1 refers); $\rho w\, g$ is the specific weight of water; Pb is the barometric pressure; P_{vt} is the vapor pressure at the time of test. The main theme of this manuscript is that there is an inverse relationship between the mass transfer coefficient and the dissolved oxygen saturation concentration, but the inverse linear relationship between K_La and C_∞^* holds only for shallow tanks, in which cases C_∞^* approaches C_S which is the solubility of oxygen in water as given by handbook values for all temperatures within the ASCE prescribed range of 10°C to 30°C. (Both Hunter's data and Vogelaar's data in Chapter 2 have shown that this temperature range can be more extensive in fact.) In fact, *the relationship between K_La and C_∞^* will have its most precise application when Z_d approaches 0, so that the saturation pressure P_S approaches 1 atm, and the saturation concentration C_∞^* approaches C_S or the oxygen solubility*, as shown by Fig. 3.5, even though, in the example cited [Yunt et al. 1988a], we only have two temperature data to go by (three if the point of origin is also counted). The author suggests that more temperatures are used to do testing in the future to verify the K_La vs. C_∞^* relationship for different tank depths, and in order to alleviate the concern about bunching up of the data at the right-hand corner of Fig. 3.4 and Fig. 3-5. In this present exercise, each tank has only one single temperature (either 16°C or 25°C) and so the author agrees that the data may not be sufficient to affirm the definite relationship between K_La and C_∞^* for different temperature, and for *shallow* tank depths only. The definition of *shallow* requires further investigations. However, the effect of gas solubility on the mass transfer coefficient can be equally examined by varying the overhead pressure rather than by varying the temperature. The case studies as given in Chapter 5 based on changing the headspace pressure from 1 atm to 3 atm dramatically illustrate the precise relationship between the baseline mass transfer coefficient and the gas solubility. The gas solubility model for oxygen as derived in Chapter 2 is thus verified.

8.1.3 Effect of Gas Flow Rate

The standard mass transfer coefficient $(K_La)_{20}$ is chiefly dependent on the average gas flow rate $(Q_a)_{20}$ and to a lesser extent also on the tank depth. Q_a is governed by another developed model as given in Chapter 2 (Eq. 2.25):

$$Q_a = Q_S \times 172.82 \times T_P \times \left[\frac{1}{P_P} + \frac{1}{P_b} \right] \tag{8.5}$$

It should be noted in passing that, in the use of the baseline model given by Eq. (7.3), the term Z_d is the diffuser depth as opposed to tank depth which is usually about 0.6 m (2 feet) above the tank bottom. The model requires an estimation of the submergence ratio (e) which can be determined by the following equation:

$$e = de/Z_d \tag{8.6}$$

The standard baseline $(K_La_0)_{20}$ has a relationship with the average gas flow rate $(Q_a)_{20}$ in water, and the relationship can be found via the following equation:

$$(K_La_0)_{20}/(Q_{a20})^q = A \tag{8.7}$$

where A is a numeric constant and q is an exponent of the gas flow rate. This constant 'A' has been defined as the *standard specific mass transfer coefficient* in this manuscript when $(K_La)_{20}$ is used in this equation. Otherwise, it is the *standard specific baseline mass transfer coefficient*, if the baseline is used, as shown in Eq. 8.7.

The baseline $(K_La_0)_T$ has a relationship with the oxygen solubility $(C_s)_T$ in water, and the relationship can be found via the following equation:

$$(K_La_0)_T/(Q_{aT})^q = B \times (1/C_{ST}) \tag{8.7a}$$

where B is another numeric constant. For the data of Yunt et al. (1988a), $q = 0.82$ (dimensionless) and $A = 0.0444$ (see Fig. 3.3); $B = 0.4031$ (see Fig. 3.5). Eq. 8.7a effectively stipulates that the *specific baseline mass transfer coefficient* at any temperature T is directly proportional to the inverse of *oxygen solubility* at the same temperature.

8.1.4   Effect of Mixing

It should be emphasized that the above derivation assumes that bubbles rise in plug flow through a tank of well-mixed water. The initial bubble-size distribution and the rate of bubble formation are assumed to be constant, and that they depend only on the specific aeration equipment. Bubble coalescence and mass transfer of gases other than nitrogen and oxygen are considered negligible. The water temperature and the ambient air temperature as well as the bubble feed gas temperature are assumed to be equal and constant. Mass transfer through the water surface at the top of the tank is neglected.

The above assumptions effectively ignore the fact that transfer devices typically produce irregularly sized bubbles that often swarm in various hydrodynamic patterns, e.g. spiral roll devices vs. full-floor coverage. In addition, oxygen transfer takes place during bubble formation, bubble retention and bubble exit at the surface. (In ordinary aeration tanks, the average hydraulic retention time for a fine bubble is in the order of 20-80 s. This reduces the time of bubble formation to a negligible fraction of the total bubble residence time. Therefore, the gas transferred in the bubble formation process is a negligible fraction of the total gas transferred [Rosso and Stenstrom 2006]). Therefore, within reasonable boundaries to be established by future research and testing on tanks of different heights, it is reasonable to assume that, statistically, these effects are more intensive in nature than extensive (i.e. these effects can be considered to be less dependent on scale than the mass transfer coefficient will be), so that they can be controlled by similitude. Calibration factors have been incorporated into the equations to account for such variables. However, these assumptions cannot be all correct, especially for the fact that the impeller speed in a sparger system may have a significant impact on the $K_L a$ value. It is assumed that, in this exercise, the rotating speed, if any, is low to moderately sufficient to maintain a continuously stirred CSTR (completely stirred and mixed tank reactor) and not affect the $K_L a$ value from tank to tank, and so the transfer rate is reasonably uniform, so that the instantaneous DO concentration throughout the tank is deemed distributed uniformly and constant.

8.1.5   Effect of Liquid Temperature

If Vogelaar et al.'s [Vogelaar et al. 2000] data is re-examined, which is based on testing in a 3-litre bottle, the exact linear relationship between $K_L a$ and the inverse of C_∞^* is confirmed (within tolerances of the experimental errors), as shown in Table (8.1) below (which is the same as Table 2.3). When $K_L a$ is plotted against the reciprocal of saturation concentration (which is similar to solubility because of its small scale as it is only a 3-litre bottle), the correlation is $R^2 = 0.9923$ (see Fig. 8.1), which is similar in terms of correlation to Fig. 3.5 in Chapter 3 using the baseline $K_L a_0$.

TABLE 8.1 Vogelaar et al. (2000) test data of $K_L a$ and C_∞^* at different temperatures at a fixed gas rate (3 vvm)

$T(°C)$	$T(K)$	$(T/1000)^5 * 10^4$	$C_\infty^* (mg/L)$	$1/C_\infty^*$	$K_L a (h^{-1})$
	0	0		0	0
20	293.15	21.65	9.19	0.1088	22.4+/-0.4
30	303.15	25.60	7.43	0.1346	26.0+/-0.1
40	313.15	30.11	6.5	0.1538	30.6+/-0.2
55	328.15	38.05	5.15	0.1942	38.8+/-1.5

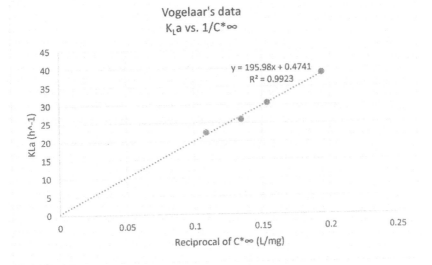

FIGURE 8.1 Plot of $K_L a$ vs. $1/C_\infty^*$ for Vogelaar's data

The y-intercept is so small it can be considered zero for all intents and purposes. This shows that a laboratory-scale test can give a similar value as the baseline calculated from a higher-scale tank. The author believes that, owing to its shallow depth, the measured K_La in this experiment is not much different from the baseline, K_La_0, so that, for all intents and purposes, this measurement of K_La can be regarded as a true measurement of the baseline K_La_0. This proves that for very shallow tanks, where C_∞^* approaches C_S, there is a definite linear correlation between the two parameters K_La and C_∞^*, for varying temperatures and under a constant volumetric gas flow rate (or average volumetric gas flow rate if the tank height becomes significant, as calculated by Eq. (8.5)). This relationship becomes less precise the further the tank height departs from the "shallow" tank criterion. This good-fit relationship cannot be simulated by the Arrhenius equation using $\Theta = 1.024$ [Lee 2017] because the temperature range is much wider in this case (0 to 55°C) than what the Θ model can handle; however, using $\Theta = 1.016$ would give a good fit as well. But, for the theta model, this is a 'chicken or egg' problem. Without the experimental data in a simulation, one would not know which Θ value to use. The 5th power model does not have this problem, and the model is good for such temperature range [Lee 2017]. With this model, any one single set of data (K_La and C_∞^*)$_T$ would be sufficient to predict any other set of data within this temperature range of 0°C to 55°C. The 5th power model was derived in Chapter 2 and is stated by Eq. 2.47 as below:

$$(K_La)_{20} = (K_La)_T \left(\frac{(\rho E\sigma)_{20}}{(\rho E\sigma)_T} \right) \left(\frac{T_{20} + 273}{(T_T + 273)} \right)^5$$

where the symbols are as defined in Chapter 2. This equation is applicable to a tank of small height. The sensitivity of this model to tank height requires further investigations.

8.2   THE SPECIFIC STANDARD BASELINE $(K_La_0)_{20}/Q_{a20}{}^q$

As for the normalization to standard temperature and pressure, on the other hand, the relationship between $(K_La_0)_{20}$ and Q_{a20} is a power curve such as is given by Fig. 3.3, where the exponent is found to be 0.82, bearing in mind that Q_a is calculated by Eq. (8.5), and the baseline K_La_0 is calculated from the measured K_La using the depth correction model. Here the relationship between the baseline and the gas flow rate is not linear and not known until the best fit curve is plotted in Fig. 3.3 with an $R^2 = 1$, giving the value of the exponent as 0.82. It is not a circular logic because, before the curve fitting the value of the exponent is unknown; it can only be determined by plotting the data to find the best correlation using curve fitting because, unlike temperature, gas flow rate is an extensive property that is dependent on both the tank height and the gas supply Q_S. The plot can be determined by using the Excel Solver or similar functional software to solve for q, and minimizing the sum of squares error. This plot is expected to be true for all tank heights, but, if the mass transfer coefficient rather than the baseline is plotted against the gas flow rate, each tank height would yield a different curve.

The power curve relationship between K_La and Q_a is supported by many researchers in the literature [Hwang and Stenstrom 1985] [Jackson and Shen 1978]. Jackson and Hoech [1977] related K_La value to a power function of the superficial air velocity, and found that the exponent q varied from 1.08 to 1.13. King [1955a, b] showed that the rate of oxygen absorption varied from 0.825^{th} to 0.86^{th} power of air flow rate depending on liquid depth and geometry. The exponent is dependent on the diffuser type and, therefore, it must be established by testing, as recommended in Chapter 5 (see Fig. 5.1 for the flow chart procedure). Zhou et al. [2012] performed testing on a full-scale wastewater treatment plant in Wuxi, China, and found that $(K_La)_{20}$ is proportional to $Q^0.877$ for fine bubble aerators. Long ago, Eckenfelder [1966] reported that the typical bubble diameter d_b is related to the air volumetric flow rate as:

$$d_b \alpha Q_a^{q'} \tag{8.8}$$

where q' is an empirical coefficient ranging from 0.2 to 1.0. Since the interfacial area per unit of tank volume V is given by Eq. 2.19 in Chapter 2:

$$a = 6Q_a \cdot Z_d/(d_b v_b V) \tag{8.9}$$

and since it is assumed that d_b is a pure function of Q_a as given by Eq. 8.8, the interfacial area is a pure function of Q_a, also. Therefore, K_La must be a pure function of Q_a also (since it is a product of K_L and a), raised to some power $q = (1 - q')$ and assuming the bubble velocity is constant for fine bubble diffusers for bubble size between 0.7 mm to 5 mm (Eq. 4.8 in Chapter 4). In light of the great variety of the exponent value in practice, the exponent is best determined by curve fitting, and in the case of the FMC diffusers, the exponent that gives the best fit to the data is 0.82. However, it is proven in this book that the exponent is independent of depth, provided that other variables are held constant. The established best-fit exponent value is then used to normalize K_La in order to produce the plots given in the main manuscript. The constancy of the exponent requires further investigation.

8.3 CLEAN WATER COMPLIANCE TESTING

This book has explained that it is possible to *design* for a system of air aeration or oxygen aeration based on testing in clean water in accordance with the ASCE 2-06 standard together with the concept of a baseline, and therefore the usefulness of the standard has been augmented enormously. It is hoped that this book would serve as a standard guideline for professional practitioners- engineers, owners, and manufacturers alike in evaluating the performance of aeration devices operating at full-scale and under process conditions. The methods presented in this book are intended for compliance testing of such, even though performance under process conditions is affected by a large number of process variables and wastewater characteristics that are not easily controlled.

Different from other textbooks, this book aims to solve a pressing engineering problem using the fundamental theories and validated by experimental results extracted from the literature. There are several major discoveries in the book:

1. It is the *height-averaged* volumetric gas flow rate that is proportional to the mass transfer coefficient, not the standard mass gas flow rate;

2. The proportionality function between mass transfer coefficient and gas flow rate is a power function but it is the *baseline* mass transfer coefficient that is exactly correlated to the mean gas flow rate, not the mass transfer coefficient itself;

3. It is the *baseline* mass transfer coefficient that bears an exact correlation to the oxygen solubility in water, not the mass transfer coefficient itself;

4. The proportionality between the *baseline* and solubility is an inverse linear function, and is in concurrence with the gas solubility law as explained in Chapter 2;

5. The mass transfer coefficient can be correlated to the equilibrium saturation concentration but only approximately, when testing under barometric pressure at the top surface and would be less accurate the further the depth increases;

6. Henry's Law is verified, and extended to create a new general gas solubility law (see Chapter 2), much like Boyle's Law is extended to form the universal gas law;

7. The new findings can be logically applied to real situations (although not yet verified by testing), such as aeration performance compliance testing in a full-scale facility in in-process wastewater treatment.

8.4 WASTEWATER COMPLIANCE TESTING AND DESIGN

Chapter 6 presents a brand-new concept about oxygen transfer in the field, and Chapter 7 gives recommendations for further research requirements on this new concept. The novel hypothesis is that the *net* oxygen transfer rate (OTR) is in fact affected by the microbial respiration rate R because of the additional resistance, produced and influenced by the microbes, that happens to be the oxygen uptake rate (OUR) at steady-state. The salient equation for calculating the OTR_f is given by Eq. 6.25 which develops into Eq. 6.32. The hypothesis was based on the argument that the oxygen transfer capacity (OTR_{ww}) of an aeration system is *not affected* by the respiration rate of micro-organisms present in the water, so that the mass transfer coefficient $K_L a_f$ is constant relative to the amount of microbes, and only varies relative to the wastewater characteristics. (OTR_{ww}) gives rise to wastewater or in-reactor solution mass transfer as opposed to the *net* transfer of the oxygen transfer rate in in-process water (OTR_{pw}). Based on the premise that the *net* oxygen transfer rate is now affected by the microbial oxygen uptake rate, it must be the vector sum of the aeration-system transfer rate and the respiration rate which is a negative quantity because of the resistance of the floc. If this resistance is not dependent on whether the system is at steady state or not, then it can be determined by a mass balance at steady state such as by adjusting the gas flow rate until a steady state is reached, since this resistance must be equal to the respiration rate at steady state. Therefore, by the principle of superposition, $(OTR_{ww}) - (OTR_{pw}) = R$. However, when the gas flow rate is altered, the DO level changes as well, and it must be assumed that this alteration of the DO would not affect the respiration rate or any other factors that may affect the value of the mass transfer coefficient, such as the mixing intensity. The logical steps for the derivation of Eq. 6.32 can be summarized as follows:

In any closed system of bulk liquid under aeration, the oxygen uptake rate by the bulk volume must be equal to the rate of oxygen transfer into the bulk liquid volume, if there are no chemical reactions involved. Therefore, considering a unit bulk volume,

$$OUR = OTR \qquad (8.10)$$

But OUR has two components: the uptake by the bulk liquid through the process of diffusion and dissolution, and the uptake by the microbial communities; therefore, by the Principle of Conservation of Mass,

$$OUR = \frac{dC}{dt} + R \qquad (8.11)$$

where dC/dt is the accumulation rate. Substituting Eq. 8.11 into Eq. 8.10, therefore,

$$OTR = \frac{dC}{dt} + R \qquad (8.12)$$

But the effective oxygen transfer in in-process water is affected by the respiration rate, so that

$$OTR_{pw} = OTR_{ww} - R_{bf} \qquad (8.13)$$

where OTR_{pw} is the oxygen transfer rate in process water; OTR_{ww} is the oxygen transfer rate in wastewater; R_{bf} is the resistance to transfer by the biological floc that provides a *negative* driving force (this resistance can be determined by the off-gas method that measures the differences in the off-gas oxygen mole fraction arising from the changes in the *gdp*). The oxygen transfer rate in the wastewater is a function of the mass transfer coefficient and the *positive* driving force given by the concentration gradient between the saturation DO concentration and the actual DO in the water. Therefore,

$$OTR_{ww} = K_L a_f (C^*_{\infty f} - C) \qquad (8.14)$$

Based on the assumption that the resistance is the same as the respiration rate, therefore, from Eq. 8.13,

$$OTR_{pw} = K_L a_f (C^*_{\infty f} - C) - R_{bf} \qquad (8.15)$$

and

$$OTR_{pw} = K_L a_f (C^*_{\infty f} - C) - R \qquad (8.16)$$

This equation is equivalent to Eq. 6.25 or Eq. 6.32. Substituting this equation into Eq. 8.12, therefore,

$$\frac{dC}{dt} = K_L a_f (C^*_{\infty f} - C) - R - R \qquad (8.17)$$

This equation is equivalent to Eq. 6.52. Hence, it is proven that the accumulation rate in the bulk liquid is given by the transfer rate *minus* twice the respiration rate of the cells within the liquid. This equation differs from the ASCE-18-96 Guidelines by a single R in the equation 2 of the Guidelines which needs to be corrected for diffused aeration. The above argument is based on a batch process, and so for a continuous process, Eq. 8.17 would need to be amended to include the wastewater flow rate and the DO concentration in the influent in such cases. The takeaway from the arguments presented in this book is that the mass transfer coefficient is affected by two major effects in in-process water oxygen transfer, namely, the wastewater characteristics, and the gas-side gas depletion rate gdp accompanying the biological floc resistance due to the microbes. All the effects are associative in nature, but the first effect is associative by scale, while the other effect is associative by addition (superposition). The current ASCE equation has treated both the effects as by scale, so that the oxygen transfer rate is given as:

$$\text{OTR}_f = \alpha K_L a(C_{\infty f}^* - C) \tag{8.18}$$

The parameter $\propto$ is a lumped parameter that includes both effects together bound into this one single parameter. The transfer equation advocated by the author is written as:

$$\text{OTR}_f = \alpha' K_L a(C_{\infty f}^* - C) - R \tag{8.19}$$

where α' is associated with the water characteristics only. Given that the oxygen accumulation rate (or the liquid phase uptake rate) in the bulk liquid is the vector mathematical sum of the oxygen transfer rate and the microbial uptake rate, this concept has resulted in a mathematical relationship given by Eq. 8.17 above.

It must be remembered that the mass transfer coefficient in clean water is determined by a non-steady state test. It cannot be determined by any steady-state test. On the other hand, when determining the mass transfer coefficient for in-process water, a steady-state or quasi-steady state test is required. Even if a non-steady state method is used, a pseudo-steady state is still required for testing in-process water. Hence, the gas depletion rates between the two types of test must be different. This difference must be reconciled if the intention is to use the clean water coefficient as a baseline for the in-process coefficient. The book has suggested that further testing is needed to validate this mass balance equation (Eq. 8.17), as explained in detail in Chapter 7.

The current use of a single constant value to represent the α-factor as exhibited in Eq. 8.18 has a tremendous flaw as is recognized by Baquero-Rodriguez et al. (2018). Their proposed solution is to change the current practice of a constant alpha (α) to using a dynamic α-factor, and to use a dynamic model to describe aeration energy demand, both in 24-hour periods with organic load variations and α-factor changes. It would be interesting to compare their results, when such a dynamic model becomes available, to the results based on Eq. 8.19 that uses the approach recommended in this book, which is to separate the dual effects of *respiration rate* and *wastewater characteristics*, and as more data with regard to both approaches are gathered. The respiration rate is a function of the organic loading rate and the amount of bacterial biomass (the respiring cells) present in the mixed liquor which is a function of the MLSS concentration. In the author's opinion, the respiration rate R is easily measurable, such as by means of the ASCE method (ASCE 1997) for *in-situ* oxygen uptake rate measurement or other methods as described in Chapter 7. However, the Guidelines ASCE 18-96 have not given an assessment of the recommended steady-state column method in detail and how well this method compares with the *ex-situ* methods which, according to the Guidelines, depend much on the time lapse between sample collection and uptake rate measurement. This book has recommended a dilution method that may give a better estimate of the *in-situ* respiration rate, as long as the test is carried out as soon as the sample is withdrawn. The author believes the depletion of substrate in the sample is much slower than the depletion rate of the dissolved oxygen, so that the rate of DO decline without aeration should give a good estimate of the microbial oxygen uptake rate. An accurate measurement of R is critical to the approach using Eq. 8.19. The conventional BOD bottle method is not recommended as it tends to over-estimate the *in-situ* respiration rate. The proposed concept of separating out the effect of respiration from other effects has resulted in a different α-factor symbolized as alpha' (α') that is to be used in this bubble-oxygen transfer rate equation (Eq. 8.19).

8.5   THE MAIN BREAKTHROUGH

The discovery of a *Standard Specific Baseline Mass Transfer Coefficient* represents a revolutionary change in the understanding, designing, operation and maintenance of the aeration equipment, as well as in providing the baseline for future research, development and design. Compliance testing means that, *subject to certain achievable constraints*, all measurements of oxygen transfer in clean water in accordance with the standard ASCE 2-06 should yield the same standardized specific baseline. Simulation means that such baseline as measured is used for scaling up and predicting performances in raw wastewater aeration through a *constant* correction factor alpha' (α') for the parameter $K_L a_0$ and through knowledge of the respiration rate R. In this book, α' is treated as dependent only on the characteristics and nature of the waste. This is substantially different from the classical method of designing the in-process mass transfer coefficient $K_L a_f$ where the parameter alpha (α) must be designed as a range. Using the transfer of oxygen to tap water as the datum, the new approach of calculating via the use of a baseline can now be used to relate the overall mass transfer coefficient of the wastewater to that of tap water. This also means that a bench-scale determination will become meaningful for translating such test results to full-scale, pending further testing and validation.

As stated in Chapter 3, this finding, if proven to be correct, may be utilized in the development of energy consumption optimization strategies for wastewater treatment plants and may also improve the accuracy of aeration models used for aeration system evaluations. The major achievement of this book was showing that gas transfer is a consistent relativistic theory of molecular interactions based on the Standard Model, and matched the predictions observed in experiment. The standard model is really the universal model that everybody is looking for, recognizing that the parameter estimation for $K_L a$ in a typical set of non-steady state, clean water oxygen-transfer data is not the real $K_L a$ for all types of aeration equipment. Therefore, solving for the real $K_L a$ is almost like an impossible task. Instead, all we can do is make some assumptions and either tease out some higher-order approximate terms or to examine the specific form of a problem and attempt to solve it either numerically such as that carried out by McGinnis et al. (2002), or mechanistically using all the physical laws, theories, and mathematics available, such as carried out in this book so that the problem becomes solvable, such as by the assumption of a constant bubble volume and constant bubble velocity.

We can then extract how the behavior of a solvable system differs from the general system in real life and find corrections by identifying the important variables in the solvable system that can be calibrated against real situations, and then apply those corrections to a more complicated system that perhaps we cannot solve. The corrected calibrated model can then form a baseline from which the standard model can be adapted to this baseline model that would yield a baseline $K_L a$ that would be true for all types of aeration equipment (see case studies presented in Chapter 5).

Hitherto, the primary challenge was the appearance of divergences in the mass transfer coefficient calculations and estimations. The whole procedure of renormalization to a baseline and to a depth-averaged gas flow rate was a great important achievement even if it had been another three decades before it was properly understood (the author first postulated the concept of gas-phase gas depletion in 1978) — these theories could have been thrown away for that interim period, possibly delaying physics for 30 to 50 years.

REFERENCES

ASCE-2-06. (2007). Measurement of Oxygen Transfer in Clean Water. Standards ASCE/EWRI. ISBN-10: 0-7844-0848-3, TD458.M42 2007.

Baquero-Rodriguez, Gustavo A., Lara-Berrero, Jaime A., Nolasco, Daniel, Rosso, Diego. (2018). A critical review of the factors affecting modeling oxygen transfer by fine-pore diffusers in activated sludge. Water Environment Research 90(5): 431.

Hwang, H.J., Stenstrom, M.K. (1985). Evaluation of fine-bubble alpha factors in near full-scale equipment. Journal WPCF, Volume 57, Number 12, U.S.A.

Jackson, M.L., Shen, C-C. (1978). Aeration and Mixing in Deep Tank Fermentation Systems. J. AIChE, 24(1): 63.

Jackson, M.L., Hoech, G.W. (1977). A Comparison of Nine Aeration Devices in a 43-Foot Deep Tank. A Report to the Northwest Pulp and Paper Association, Univ. of Idaho, Moscow.

King, H.R. (1955a). Mechanics of oxygen absorption in spiral flow aeration tanks: I. Derivation of Formulas. Sewage and Industrial Wastes 27: 894.

King, H.R. (1955b). Mechanics of oxygen absorption in spiral flow aeration tanks: II. Experimental work. Sewage and Industrial Wastes 27: 1007.

Lee, J. (2017). Development of a model to determine mass transfer coefficient and oxygen solubility in bioreactors, Heliyon, Volume 3, Issue 2, February 2017, e00248, ISSN 2405-8440, http://doi.org/10.1016/j.heliyon.2017.e00248.

Lewis, W.K., Whitman, W.G. (1924). Principles of gas absorption. Industrial and Engineering Chemistry 16(12): 1215-1220. Publication Date: December 1924 (Article) DOI: 10.1021/ie50180a002.

Rosso, Diego, Stenstrom Michael K. (2006). Alpha Factors in Full-scale Wastewater Aeration Systems. Water Environment Foundation.

Vogelaar, J.C.T., KLapwijk, A., Van Lier, J.B., Rulkens, W.H. (2000). Temperature effects on the oxygen transfer rate between 20 and 55 C. Water Research 34(3): 1037-1041.

Yunt, Fred W., Hancuff, Tim O., Brenner, Richard C. (1988a). Aeration equipment evaluation. Phase 1: Clean water test results. Los Angeles County Sanitation District, Los Angeles, CA. Municipal Environmental Research Laboratory Office of Research and Development, U.S. EPA, Cincinnati, OH.

Zhou, Xiaohong, et al. (2012). Evaluation of oxygen transfer parameters of fine-bubble aeration system in plug flow aeration tank of wastewater treatment plant. Journal of Environmental Sciences 25(2): 295-301.

Index